HOME
SECURITY

arl Hammer

HOME SECURITY

How to Select Reliable Locks and Alarms for Your Home, Office, or Car

Paladin Press • Boulder, Colorado

Home Security:
How to Select Reliable Locks and Alarms for Your Home, Office, or Car
by Carl Hammer

Copyright © 2001 by Carl Hammer

ISBN 1-58160-121-2
Printed in the United States of America

Published by Paladin Press, a division of
Paladin Enterprises, Inc.
Gunbarrel Tech Center
7077 Winchester Circle
Boulder, Colorado 80301 USA
+1.303.443.7250

Direct inquiries and/or orders to the above address.

PALADIN, PALADIN PRESS, and the "horse head" design
are trademarks belonging to Paladin Enterprises and
registered in United States Patent and Trademark Office.

All rights reserved. Except for use in a review, no
portion of this book may be reproduced in any form
without the express written permission of the publisher.

Neither the author nor the publisher assumes
any responsibility for the use or misuse of
information contained in this book.

Visit our Web site at www.paladin-press.com

TABLE OF Contents

List of Illustrations IX

Preface XV
    Why You Should Read This Book
    Burglars Kill
    Every Security System Can Be Breached
        but Not by Any Burglar
    What You Will Learn from This Book

Chapter 1: How the Burglar Breaks into Your Home 1
    The Burglar: Who Is He?
    The Burglar's Tactics and Techniques
    The Average Burglar at Work
    Sophisticated Burglary
    Common Mistakes That Give the Burglar an Easy Ride
    File Cabinets and Desk Drawers

Chapter 2: How to Select a Reliable Alarm System 53
    The Security System's Three Lines of Defense
    How the Alarm System Functions

Chapter 3: Alarm Sensors 89
    Magnetic Reed Switches and Wire Contact Systems
    Window Foil
    Windowpane-Mounted Glass-Breakage Detectors
    Vibration Detectors and Inertia Sensors
    Infrasound Detectors
    Field-Effect Sensors
    Sound Detectors and Heat Detectors
    Pressure Mats, Plunger Switches, and Contact Strips
    Ionization Detectors
    Photoelectric Cells and Invisible Beam Detectors
    Passive Infrared Detectors
    Microwave Motion Detectors
    Ultrasonic Motion Detectors
    Visible-Light Detectors
    Video Detectors
    Barrier Sensors and Analyzers

Chapter 4: Personal Attack Alarms 131

Chapter 5: Fire Alarm Systems 135

Chapter 6: The Most Reliable Locks—
           and How Intruders Pick Them 141
    Warded Locks
    Lever Tumbler Locks
    Disk Tumbler Locks
    Pin Tumbler Cylinder Locks
    Tubular Cylinder Locks
    Magnetic Locks
    Combination Locks
    Master Key Systems
    Common Lock-Picking Tools

Chapter 7: Other Common Locks .185
    Simple Suitcase Locks
    Padlocks
    Briefcase and Padlock Combination Locks

Chapter 8: Reinforced Doors and Security Bars 193
    Reinforced Doors
    Security Bars

Chapter 9: How to Protect Your Safe 197
    Safes and How Burglars Crack Them
    Combination Locks

Chapter 10: Gun Storage 211

Chapter 11: The Small Office or Shop 213
    Shoplifter Detection Systems
    Smoke Generators

Chapter 12: Vehicle Security 219
    Vehicle Locks
    Vehicle Doors, Windows, and Trunks
    Car Alarm Systems

Chapter 13: Personal Safety at Home 241

Chapter 14: Security Away from Home 245

This book is dedicated to all those ordinary men and women who have suffered the trauma of having their homes burglarized and their belongings stolen or vandalized. It is my sincere hope that the publication of this book will make the job of the common burglar (as well as his sophisticated counterpart) both more difficult and dangerous.

LIST OF Illustrations

1. Hinge bolt
2. Mortise rack bolt
3. Barrel bolt
4. Door limiter
5. Unlocking a door chain with a rubber band and a tack
6. Unlocking a door chain with a rubber band and a bent coat hanger
7. Opening a door with the Z-wire
8. Transom entry
9. Louver window lock
10. Opening an unlocked sash window from the outside
11. Opening a bolted sash window from the outside

12. The cam of the plug operates the bolt in most file cabinets.
13. Pulling down the bolt with a jimmy
14. Forcing the bolt mechanism down to release the drawer catch
15. The drawer catch passes the bolt mechanism, riding on an inserted strip of steel.
16. Sliding cabinet door plunger lock
17. The difference between normally opening (NO) and normally closed (NC) switches
18. Key switch
19. Wiring switches in series (left) and parallel (right)
20. Electromagnetic lock
21. Lock switch installed in mortise lock
22. The jumper cable terminal and its position on the door frame
23. Magnetic reed switch
24. NO and NC magnetic reed switches
25. Recessed magnetic reed switch
26. Wire contacts
27. Window foil
28. Glass-breakage detector
29. Vibration detector
30. Field-effect sensors
31. Sound detectors mounted in a bank vault
32. Pressure mat
33. Plunger switch
34. Contact strip bridged with jumper cord
35. Photo relay sensor (the detector, left, and the reflector, right)

36. Infrared beams arranged for perimeter protection
37. The use of IR reflectors to protect a corridor
38. The detection zones of a passive IR detector
39. The passive IR detector's sensitivity to movement across its segments, rather than toward or away from it
40. Microwave motion detector and its detection zone
41. The microwave motion detector's greatest sensitivity to movement toward or away from it, with sensitivity to movement in another room as well
42. The ultrasonic motion detector: also most sensitive to movement directly toward or away from the sensor
43. Inertia barrier sensor used on a fence
44. Microwave fence
45. Geophone
46. Smoke detector
47. Thermal detector
48. Ancient Egyptian iron keys
49. Rim lock (left) and mortise lock (right)
50. Cylinder lock
51. Lever lock
52. Ancient Roman key
53. Side ward
54. Another type of side ward, and end wards, the latter milled on both ends
55. Single-headed and double-headed skeleton keys, seen from the edge (right) and from the side (left)
56. Chubb lock with key
57. Lever tumbler lock
58. Picking the lever tumbler lock
59. Disk tumbler lock

60. The five disk tumbler variations and their positions in the keyway
61. Simple disk tumbler lock
62. Double-sided disk tumbler lock
63. Yale lock with key inserted
64. Pin tumbler cylinder lock
65. High-security drivers: mushroom (left) and spool (right)
66. Raising a pin to the shear line
67. Protective inserts on the face of high-security cylinders
68. Tubular cylinder lock
69. Picking the tubular cylinder lock
70. Tubular lock-picking tool
71. Magnetic padlock
72. Pushbutton combination locks (left) and a digital keypad lock (right)
73. Disk tumbler prepared for master keying
74. The master key, relying on raising one or more master pin tumblers to a "new" shear line
75. An example of a master key hierarchy
76. Picking a pin tumbler cylinder lock with a safety pin and a small screwdriver
77. Various types of lockpicks
78. Lockpicks for warded locks
79. Torque wrenches
80. Lockpicks for double-sided disk tumbler locks
81. L-shaped lever lock pick, the same pick improvised from wire, and a suitable torque wrench
82. Pick gun
83. Snap pick
84. Picking a suitcase lock

85. Warded padlock, with the internal wards illustrated, and a suitable skeleton key
86. T-shaped wire pick
87. Sesame combination lock
88. Sesame padlock
89. Manipulating the sesame padlock
90. Sliding grille
91. Floor safe
92. Floor safe in a timber floor
93. Wall safe
94. Small wall safe disguised as electrical power socket
95. Dial combination lock
96. Wheel with six false gates
97. The Securitag system
98. Sidebar lock
99. The coat hanger bent in a loop or triangle
100. The paint scraper forced between the edge of the window and weatherstripping
101. A loop of nylon line fixed to the slotted cleaning attachment
102. Tools for opening rear ventilation windows with locking buttons and how they are used
103. Opening the front ventilation window and tools for this purpose
104. Removing the cylinder from the outside with the L-shaped tool
105. Pin-switch and its location in the car
106. Microtransducer
107. Steering wheel hook lock

Preface

We have all watched the scene on TV countless times. The happy, hard-working family, peacefully at sleep at home. Suddenly the bedroom door is thrown open, a gang of violent thugs, escaped convicts, you name it, rushes into the room. Then terror, rape, murder. Another family dead, another home shattered.

Could it happen to you? Yes, it could. Such scenes do take place in real life all too often.

WHY YOU SHOULD READ THIS BOOK

So how can you protect yourself? You may think the solution is to buy a gun and leave it loaded next to your bed. But picture that scene again. Will you really wake up in time? Will you have the presence of mind, in the middle of the night, to find your gun, aim it, and actually hit and kill all the assailants before they do you in? They were awake; you were not. They are accustomed to violence; you are not. Unless you are a professional soldier, home fresh from a war zone, forget it. That bedside gun will not protect you.

For those of you who think you are good enough with a gun to take out the assailants before they can harm you or your family, remember that in the past 10 years, no fewer than 29 law enforcement officers were killed during burglaries in progress. And they were not asleep in the house when the burglar entered. The only time your bedside gun is likely to be used, and sad statistics confirm this, is when your child uses it to play with, with possibly fatal consequences.

Do you find this introduction depressing? You should, because these facts are real. But all is not lost. There are ways, easily accessible to all and not too expensive, to fortify and secure your home so that no intruders will ever be able to enter the building, let alone your bedroom, until long after you have become aware of them . . . if they are able to enter at all. The solution? To secure your external doors and windows so that burglars cannot enter easily and to install an efficient alarm system that will warn you if they still try. And this book will teach you how.

BURGLARS KILL

Far too many burglars have a far too easy time getting into homes. Why do so many people leave their houses so easy to break into, even if the home is full of valuables and important family mementoes? And this in an age when many burglars do not care if they find the owners at home after they have broken into it—and then far too often solve the predicament by killing or maiming those whom they do find.

According to data from the U.S. Department of Justice, 32.8 percent of Americans keep a gun in the house as a precaution against crime. But only 19.5 percent use a security system to guard themselves and their families. Yet a reliable security system is far more likely to detect a burglary than an absent or sleeping man, armed or not.

Burglary is no mere inconvenience. Not only is the psychological shock of seeing your home destroyed and your belongings stolen considerable, the chance of actually being killed at home by an unscrupulous and frequently drug-crazed burglar is far too high. And yet many choose to ignore this very real peril, preferring to think that this could only happen to others.

The experience of having your privacy and home violated by total strangers can cause severe trauma to its victims, even if they were not present. In many cases, people have had to move from a house they have lived in for years, a home they loved, because of the traumatic effect that a burglary had on them. Why? Because they no longer feel at ease in their own homes. Insurance may help to replace the occasional television, but no insurance will ever replace your far more treasured personal items. No insurance will cover the loss of safety and feeling of vulnerability that you must confront in your own home.

There are ways to protect yourself that are neither too expensive nor time consuming. However, knowledge will be your best asset in the war against burglary. To fortify your home, you must apply proven methods and technology. Home security may be a choice between life or death for you and your family. Is it not then unbelievable that so many present-day alarm systems, including numerous American do-it-yourself security systems, are characterized by an unfathomable naivete? Although the electronics might be highly advanced and true state of the art, the actual application of the system is far too often easy to defeat. Why? Because of the extremely large number of companies that wish to cash in on the lucrative security market, each trying to attract attention by the latest gimmick, whether it works or not under actual field conditions.

Any salesman of alarm and other security systems worth his salt will be eager to point out how efficient his particular brand is. He will not mention the weak points of whatever system he is marketing. And be in no doubt of this: every alarm system can be bypassed. Every security door or security lock can be broken. Every safe can be breached. Some systems are easier to penetrate than others, and it is your job to know what is reasonably safe and what is not. The salesman will only do the selling. Do not rely on him. Besides there have been cases when salesmen or their associates sold information about their customers to the very burglars the customers were trying to keep out. When it comes to the security of your home, you should preferably trust no one but yourself.

EVERY SECURITY SYSTEM CAN BE BREACHED BUT NOT BY ANY BURGLAR

If every security system can be breached, you may ask, what is the point of installing one? The answer is simple but perhaps not obvious. The prime purpose of your security system is not to capture intruders but to convince any potential burglar casing your house that a break-in will be too difficult, take too long, and generally be too dangerous to attempt. Why break into your home, when most of the neighboring houses remain unprotected? An efficient security system will discourage burglary attempts by ordinary criminals. The true professionals may find a way in, but unless your art collection rivals that of the Metropolitan Museum of Art, you can live with this risk. Such men are not interested in ordinary homes. Even if you would be at risk from professional burglars, you can greatly reduce the risk by following a few reliable guidelines on how to select and install your security system.

If you have the necessary knowledge, cheap alarm system components bought in your local electronics or hardware shop can produce a security system just as secure, if not more so, than what the security companies in your neighborhood will try to sell you at greatly inflated prices. It is not really the components that count but the knowledge of the designer of the system. After reading this book, you will no doubt know just as much, if not more, about a particular system's strong and weak points that any so-called security system salesman. The only thing that you will need, in addition to the information in this book, is the technical specifications on the systems available in your particular area. Whatever brand you finally acquire, the general guidelines in this book will help you secure your home. Who knows, after installing your own system, you may even consider entering the security system business yourself as a lucrative sideline, especially if you already have some experience as an electrician or construction worker.

WHAT YOU WILL LEARN FROM THIS BOOK

The first part of this book explains how burglars break into homes and offices. The full range of burglars is covered, from the run-of-the-mill burglar to the most sophisticated art thief and government agent.

Next comes a major section on alarm systems and alarm sensors. The best way to install a safe system is to plan for its potential weaknesses. The text explains how common types of alarm systems can be circumvented (e.g., defeated by avoiding all sensors and switches; how the system can be defeated by other means, such as frustrating it). These two methods, circumvention and counteraction, are detailed as regards most types of today's alarm systems, as well as a few emerging types we will probably see more of in the future.

After this section is an in-depth look at locks and efficient methods to prevent burglary by reinforcing doors, windows, and so on.

The last part of the book details vehicle alarms, particular problems for shopkeepers and how they can be solved, and other special situations, including how you can protect yourself and your family members.

In the future, other access control systems might turn out to be more common than the present leading ones. Devices for identifying people through finger- and handprints, as well as voice identification systems, are currently being developed. Some of these systems are in fact already in limited use. Despite this, it is quite certain that the present alarm and locking devices will remain for a very long time indeed. After all, the present locking technology for the most part relies on and dates back to the 19th century.

It is likely, however, that there will be no clear-cut borderline between future alarm and locking devices, because these two groups are approaching each other a little more for each day. For instance, this can be seen clearly in the wide variety of electronic locks and card locks. I expect that we will see more of this integration in the future. Integrated alarm and locking systems are detailed in the appropriate chapters.

In a previous book (*Expedient B & E: Tactics and Techniques for Bypassing Alarms and Defeating Locks,* also from Paladin Press), I detailed the methods used by government agents to break into and search the homes or offices of suspected enemy spies and traitors—often clandestinely and without leaving any traces. Whatever one thinks of the motives of those professionals, we have to accept that their knowledge of breaking-and-entering techniques is unrivaled. Although the vast majority of burglars rely on simple brute force, some of today's sophisticated criminals use the same methods as the government agents. So let us secure our homes so that common and sophisticated intruders alike will be deterred from attempting a break-in. The information in this book will take those people concerned about home security far into the new millennium.

NOTE: Because the author lives in Europe, all measurements follow the international metric system. The following chart should help you convert metric into U.S. measurements:

1 millimeter = 0.039 inches
1 centimeter = 0.39 inches
1 meter = 39.37 inches
10 meters = 32.81 feet

Just remember that no measurements are absolute: they must be adapted to the specifics of the actual lock or device you are working with.

CHAPTER 1

How the Burglar Breaks into Your Home

Data from the U.S. Department of Justice show that nearly every American will suffer from personal theft at least once and that 87 percent will be victimized three or more times. Theft without forcible entry will occur in nine of ten homes. Urban households have a 93 percent chance of being burglarized and rural homes 82 percent.

THE BURGLAR: WHO IS HE?

The vast majority of all burglaries are carried out by the opportunist thief who is likely to live a short distance from the house he targets. He is probably a male, and often a juvenile aged 14 to 17. This means that there is high chance that your home will be vandalized in addition to being burglarized. Vandalism includes using your living room as a toilet and starting a fire in your bedroom, so do not underestimate the loss, monetary and emotional, from it. Professional burglars certainly destroy furniture and make a considerable mess, but they want to get in and out of the premises as fast as possible.

As many as 20 percent of reported burglaries each year are committed without forcible entry. The burglar simply finds an unlocked or open door or window and walks or climbs in. Much crime can be prevented by merely installing, and using, a reliable lock.

Between 80 and 90 percent of burglars have no particular target in mind when they set out. They survey a number of homes and pick any easy target. Most burglaries are committed in 10 minutes or less. Before entering, while walking along the street, the burglar will have identified your home as probably unoccupied. He knocks at the door, and if there is no reply, he enters through an open window or an unlocked window that is easily smashed, or by breaking through the door. He starts his work at the top of the house, going through the main bedroom first to take any cash or jewelry in sight. He empties all the drawers; searches the shelves and the pockets of clothing in the wardrobes; and checks under the bed, mattress, and rug. He takes any camera equipment, expensive clocks, and other valuables, bringing them along in his pockets or a suitcase or bag that he finds in the house. After searching the other bedrooms, he will move downstairs to empty drawers and shelves in the living room and other rooms. He is even likely to search the kitchen. Finally, he will leave by the back door if the key is available or by the same way that he entered. In his trail, he will leave a considerable mess because he wants to get out fast. And, remember, the whole episode will have taken no more than 10 minutes.

THE BURGLAR'S TACTICS AND TECHNIQUES

For the intruder, the outcome of his burglary attempt depends on two important factors: tactics and techniques.

Not all burglars are sufficiently articulate to describe the two concepts in those words, instead relying on an instinctive understanding of them. You, however, need to realize the difference between tactics and techniques because they must be countered in different ways. The chosen tactics determine how to conduct the break-in and how to avoid being detected or caught. The techniques are the actual methods used for bypassing locks, doors, and other obstacles, such as alarm systems. By having a thorough understanding of the tactics used by burglars, you will avoid silly mistakes that would leave an otherwise sophisticated security system inoperative. By knowing the details of the actual techniques in use among burglars, you will better understand the technical aspects of your security system, its advantages and disadvantages.

The first thing a burglar will do, be he drug-crazed or sophisticated, is to reconnoiter the target (i.e., the actual building to be entered) and the area around the target. In military parlance, this is the reconnaissance phase. It may be as simple as strolling along the street, jimmy or crowbar in a coat pocket, looking for a target of opportunity; or it may be full-scale reconnaissance operation lasting days or even weeks in the case of sophisticated intruders targeting a major art museum or bank. In the first case, your job is simply to ensure that your home presents an impregnable front. The opportunist burglar should see it as a difficult job and move on to easier targets. If so, your security system has saved you without your ever knowing. In case of sophisticated intruders who know what they are looking for (e.g., large numbers of guns, valuable art objects), the situation is more complicated. They will already have ascertained what treasures you keep at home. Now your security system will be tested as never before. Your defenses must again show an impregnable front, without obvious weaknesses to be taken advantage of. A multilayer security system of the type discussed in this book will discourage even the most ruthless intruder. Even professionals may be discouraged and turn elsewhere.

The vast majority of all burglaries is committed by the

average opportunist, relying on brute force. Therefore, it is appropriate that we investigate his tactics first.

THE AVERAGE BURGLAR AT WORK

The run-of-the-mill burglar relies on brute force and speed. Why bother to pick a lock when it is easier and faster to break a lock or an entire door? Alarm systems are to be evaded if possible or else disregarded. Speed and force are far more important than surreptitiousness. Few police forces have the resources or time to solve a burglary, which they regard as a minor crime. As long as the burglar has escaped before anybody had time to respond to the sounded alarm, the chance that he will ever be caught is minimal.

The presence of an alarm system, if merely designed to alert a distant security company or a police station, is often completely disregarded by the intruder. The reason is that the alerted security team or police will not proceed fast enough to reach the penetrated area in time to catch the burglar, who has by then already disappeared. Furthermore, both police and security companies are reluctant to rush to a site, when the alarm system alerts them. The reason for this seemingly contradictory behavior is the fact that the majority of all alarm calls are false alarms, caused by defective alarm systems that for purely technical reasons sound the alarm even if no intruder is present.

The burglars, working alone or as a team, are equipped with anything from crowbars and sledge hammers to a carborundum wheel with circular saw attachment. The latter is an extremely hard silicon carbide grinding wheel, excellent for breaking through all sorts of hard steel and other strong materials, and thus very useful in cracking safes. Explosives might also be required, although this is less common now. In the past, explosives were commonly used for safe cracking, but a carborundum wheel is much more efficient and easier to control. Batteries for high-voltage power might also be necessary if the fuses in the building blow because of an excessive power load or if electrical power is unavailable for some other reason.

The burglar knows that the normal means of entry (e.g., the front door) is frequently more difficult to break than some other part of the building, such as a roof or a wall. The door is often reinforced, which very seldom is the case with other parts of the building, at least in smaller houses and villas. Even if the burglar goes for your front door, the actual lock is usually not his target. The other components of the lockset and the door are generally much weaker, for instance the striking plate, the hinges, the door frame, and the panels in the door. Even the wall next to the door might be weaker and thus more vulnerable than the door itself. The same might apply for the roof or ceiling or even the floor. In most buildings built in a warm climate, the walls lack insulating materials. The wall then often consists only of an empty shell that is extremely easy to break through with heavy-duty tools. Floors and ceilings are notoriously weak in most countries, whether the climate is warm or cold.

All these facts work in favor of the burglar. The level of force employed in the break-in is highly variable, of course, depending on the circumstances. Sometimes a low degree of force is sufficient. In a few cases, however, the door or the wall area around the main door was smashed through by ramming the building with a car or a heavy truck. At other times, a tractor was used for the same purpose. These methods are very efficient and quick, if somewhat noisy, approaches to burglary. Less violence might be sufficient, but remember that the average burglar will choose the most expeditious method, whatever the carnage.

Vulnerable Entry Points and Simple Precautions You Can Take

The lesson is that you must deny the burglar any weak entry points into your home. You will have to go through your home to evaluate its weak and strong points. Sometimes the structure will be strong enough that only a few weak spots (e.g., doors, windows) need to be reinforced. In many modern apartment buildings, especially above the ground floor, the only access point you need to think about is the front door. It is extremely uncommon for a burglar to break into upper floor apartments in high-rise buildings at any point other than through the front door.

In some buildings, the front door might be very difficult to break through, but a garage will be attached to the side of the house. Inside the garage there will then be only a lightly protected internal door linking the garage to the home. This door might well turn out to be a much easier target than the front door. In addition, a burglar working inside your garage is far less likely to be noticed by your neighbors, even if he makes plenty of noise. No burglar is too stupid to walk around a house he plans to break into to find its weak spots. You had better do this survey (as well as remedy any problems) before the burglar does.

In later chapters, we will detail vulnerable points and how these can be taken advantage of by burglars. What you should be looking for include trees that can be used to gain access to upper floor windows and perhaps even the roof, drainpipes that can be used for climbing, ladders left where opportunist burglars can find and use them, operable ground floor windows without locks, unlocked garage doors, trashcans or Dumpsters that can be climbed up on, and so on. Is your garage easy to break into, and can a burglar find there all the tools he needs to break into your home? When you survey your house for security, attempt to think like a burglar. How would *you* break into the house? Remember that even small windows may be vulnerable; some burglars are only children. High hedges may provide cover for a burglar to work unnoticed by your neighbors. On the other hand, burglars do not like such prickly plants as roses or holly. A row of such plants in front of the windows below the sill level serves as a further deterrent to intruders.

Some vulnerable points are easy to remedy. You can keep your ladder locked in your garage or at least chained to a fence or wall with a padlock. Drainpipes can be coated with anticlimb paint, a thick paint that dries on the outside but remains wet underneath its skin, makes the pipe greasy and difficult to climb, and leaves a sticky deposit on the climber's hands (do not coat the bottom 2.5 meters or so in case family or visitors lean against the drainpipe). Another disadvantage is that insects tend to get stuck to the paint, which can be unsightly. Metal drainpipes can also be replaced with plastic ones, because the latter are not strong enough to support a climber. You could even place a spiked metal collar around each drainpipe about 2.5 meters up. Heavy-duty tools can be kept in the house rather than in the garage. A brick wall that is easy to climb can be made a more difficult obstacle by making the top course of bricks castellated or pointed to increase the difficulty of creating a handhold. You can even place broken glass or barbed wire there to further discourage intruders, but this would be illegal in some jurisdictions. Many windows can be protected by security bars. Other possible entry points will have to be protected by more technical means, generally locks and alarm systems.

You should also consider positioning any valuable computers, stereo systems, and similar equipment so that they or the illuminated faces of the equipment are not easily spotted from outside. At least keep them turned away from the windows so they are not visible at night.

A couple of other important precautions should not be dismissed because they really deter crime, even though they are beyond the scope of this book. If there is a neighborhood watch in your area, attend the meetings. The easiest way to fight burglary is to agree with your neighbors to watch out for each other's safety. Also remember that burglars need to sell what they steal. Make it difficult for them by engraving security identification numbers on all your valuables. More information on these procedures can often be obtained from your local police station.

Doors

The door is often the weak link in the protection of the building. The simple fact that the door has to be opened frequently ensures that it cannot be made completely safe. It is, however, important to determine which is the weakest part of the door. If the door is massive, then the lock will be the weakest part, but it is much more common that the door itself is of a weaker construction than the lock installed in it. Far too many people install expensive high-security locks in fundamentally weak doors only because some salesman has lulled them into a feeling of false security. Most doors can be smashed one way or another, and, if not, the burglar has a wide range of techniques to open locked doors.

However, if a lock is installed in the wrong way or simply installed carelessly, it may be easier to break through the locking mechanism. Look out for major gaps between the door and its frame. If any of these gaps is large enough, a burglar may be able to open the door simply by inserting some suitable object or prying tool through it and wrenching the door open.

A cylinder lock is incorrectly installed if the cylinder protrudes more than 2 millimeters. If this is so, then the lock can be forced open by either pulling out the cylinder with heavy-duty tongs or drilling through the now exposed weaker side of the cylinder to destroy the locking mechanism.

A lock can almost always be drilled open. In a very few cases, the lock can then be replaced without the owner's noticing the fact, at least if the lock is a cylinder lock. If only the cylinder plug is drilled, the cylinder itself can be saved. Only the inner core needs to be replaced. However, it is not possible to reconstruct the key afterward. The core has to be a new one, although the external appearance will be the same. Obviously, the average burglar would not bother with such fine details. When drilling open a cylinder lock, the lower sets of pins just below the shear line (the full technical description follows in a later chapter) will be destroyed, and this allows the plug to be turned, because the upper pins can be kept above the shear line with a wire inserted into the lock. There are also various other ways of drilling open a lock, but these methods will destroy parts of the lockset that are exposed and clearly visible, making the intrusion extremely conspicuous. The burglar typically finds it easier to wreck the entire door.

The area around the lock is also a good choice for breaking open. If, for instance, two mortise locks are installed in the same door, the locks should be at least 40 centimeters apart or the door structure will be significantly weakened. Even though the locks might be strong, the door is then easy to break.

In other cases, the screws attaching the lock to the door might be visible and even possible to remove from the outside. This is another major construction error that makes it easy to force the door open.

A good idea is to reinforce your door(s) by fitting steel plates on the most exposed parts. If these plates are thick enough and the frame is not significantly weaker, this will cause problems for most burglars. A carborundum wheel might be necessary to cut through the steel, and to use such a power tool, the intruder needs a source of electricity. Few burglars carry powerful batteries with them. However, do not leave a functioning power socket in your garage or outside your house if you expect a burglar to carry power tools. If so, make sure that the power can be—and is—disconnected from a point inside the main building. You can always disconnect the fuse to this particular socket.

Doors with rebated wooden panels and most other types of paneled doors are very weak and can easily be broken by a hefty kick or a hacksaw blade. The burglar aims for the panel next to the lock, so he can reach in with his hand and unlock the door. It may be possible to reinforce these doors by steel sheeting on the inside rather than on the outside. If this is the case, a burglar must treat the door as a steel door, despite its inviting exterior appearance. He will very likely give up when he encounters the steel sheet, but, of course, by then the door will be damaged.

Some front doors have very flimsy lower panels. They are not only easy to break in through, but they can also be used to push out larger objects from the building. This is a technique often employed by burglars. You can reinforce these panels on the inside by stronger and thicker panels of wood, although such reinforcements can sometimes be detected from outside by the appearance of a number of screws. Wood reinforcements, unlike steel reinforcements, are possible to break through with ordinary tools, but the job takes time and creates noise. Most burglars will give up and move on.

Patio doors and sliding doors, especially those with either a wooden or an aluminum frame, are usually easy to break through by simply lifting the doors out of the fitting. A patio door lock is a security device designed to bolt the sliding door to its frame. Such a lock is never strong enough. Some patio doors are even made of plastic. Needless to say, they cannot be trusted to withstand any attempt to break in.

Glazed doors are extremely vulnerable. The intruder does not need to break a big hole, however. A small hole, enough to allow him to reach in and open the lock, is quite sufficient. A small area of glass is more easily broken—in a safe way—than larger areas of glass. If, for instance, the door is a French door, it is likely to open outward, and the latch can be reached and opened easily by breaking one of the small panes of glass in the door. Such a small pane can be broken quietly, without risking the neighbors' hearing it. Burglars do not like to cause unnecessary noise, true, but they are far more afraid of cutting themselves while breaking glass windows, so most burglars avoid entering through glass doors and windows. However, they will not hesitate to simply reach in to open the lock from the inside.

Most doors with a mail slot are very vulnerable to an attack through this opening. The intruder will simply put a crowbar through the mail slot and break open the part below. This job takes only a few seconds.

The latch can sometimes also be reached through the

mail slot itself. Special equipment can be made for this purpose, and some burglars still use it.

No door is stronger than its frame. A softwood frame is easily broken. Exposed hinge pins can also be worked on to open the door.

If a wedge had not been fitted between the frame and the wall opposite the lock during the construction of the house, then the frame is significantly weakened. In this case, the burglar can simply bend the door open by spreading the door frame, thus allowing the door to be opened without bothering with the lock mechanism. This can be done unless a very long bolt is used or the frame is very rigid. This is true even of most steel frames, at least those no thicker than 2 millimeters. Since damage to the door is a frequent outcome of this technique's use, the professional may wish to avoid this result. Ordinary burglars do not care.

Most outward-opening doors can be opened by knocking out the hinge pins and then prying the door open on the hinge side. Alternatively, the hinges can be sawed off if the pins cannot be knocked out. A strong door may also be protected by hinge bolts (Figure 1) fitted in the hinge edge of the door close to the hinge positions. These are studs set into the hinge edge of the door that engage into corresponding sockets, or recesses, in the frame whenever the door is closed. The door cannot then be forced open or lifted simply by sawing off the exposed hinges. Such a door can only be forced by either removing the lock or the frame area around the hinges and the hinge bolts. Hinge bolts are always installed in pairs, one usually below the top hinge and the other above the bottom hinge. As long as your door is fundamentally strong, the installation of hinge bolts is strongly recommended. You can do it yourself quite easily without ruining the appearance of the door, since the bolts will be visible only when the door is open.

Mortise rack bolts are sometimes used for a similar purpose, as a bolt for an external door. Such bolts are fitted at the top and bottom of a door and can be locked from the inside with a universal splined key (Figure 2). The lockable bolt is the type most commonly used today. These bolts can only be locked from the inside. The mortise bolt is a dead bolt, so it cannot be released without a key.

It is worth mentioning that in older houses some doors and French windows are usually secured with an espagnolette bolt that extends the full length of the door and consists of two vertical sliding bolts, one covering the top half of the door or window and the other covering the bottom half. Both are operated by a central handle, which is often lockable. This works according to roughly the same system and is used to keep the door in place, even if the hinges are destroyed.

Barrel bolts are used in many buildings for a similar purpose. These bolts are screwed onto the surface of the door and usually shoot into a staple on the door frame. They are usually not fit-

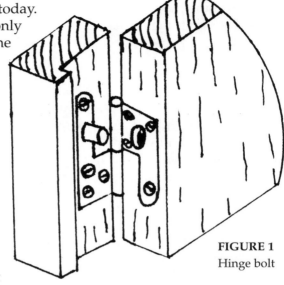

FIGURE 1
Hinge bolt

FIGURE 2
Mortise rack bolt

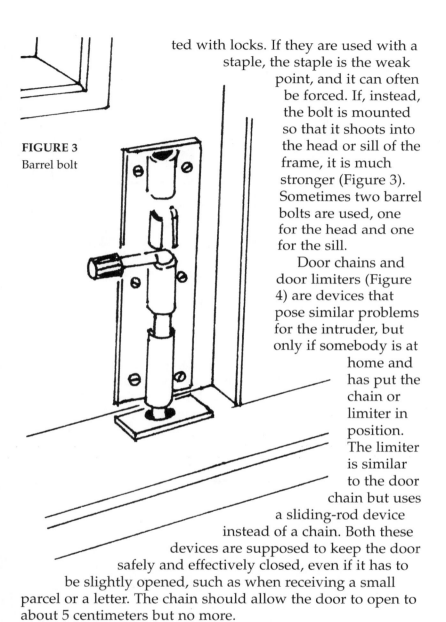

FIGURE 3
Barrel bolt

ted with locks. If they are used with a staple, the staple is the weak point, and it can often be forced. If, instead, the bolt is mounted so that it shoots into the head or sill of the frame, it is much stronger (Figure 3). Sometimes two barrel bolts are used, one for the head and one for the sill.

Door chains and door limiters (Figure 4) are devices that pose similar problems for the intruder, but only if somebody is at home and has put the chain or limiter in position. The limiter is similar to the door chain but uses a sliding-rod device instead of a chain. Both these devices are supposed to keep the door safely and effectively closed, even if it has to be slightly opened, such as when receiving a small parcel or a letter. The chain should allow the door to open to about 5 centimeters but no more.

Certain door chains do have a key-operated lockable staple on the door frame that allows the chain to be used even

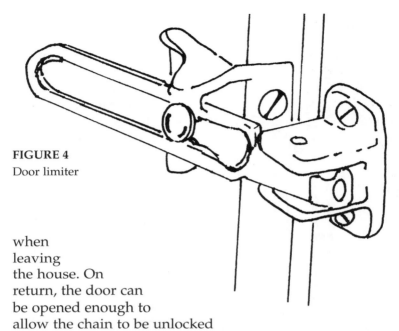

FIGURE 4
Door limiter

when leaving the house. On return, the door can be opened enough to allow the chain to be unlocked from the staple. These door chains can cause problems for the burglar, unless the staple is weak. When the burglar has managed to open the door, he will find that the door chain prevents the door from fully opening and that he cannot reach the chain to remove it. Often the chain can be forced; few door chains are sufficiently strong. There have been cases when door chains have been broken by merely opening the door, without first removing the chain. Do not rely on door chains alone.

If the door chain is too strong to break, the chain can be removed with a rubber band. The intruder reaches inside and sticks a tack in the door behind the chain assembly. He then attaches one end of the rubber band to the tack and the other end to the end of the chain (Figure 5). After making certain that the rubber band is taut, he closes the door, taking care not to lock it again. If this is difficult, he will secure the lock with adhesive tape before he closes the door, so the mechanism cannot work. When the door is closed, the rubber band will pull the chain back. If this fails to pull the chain off the slide completely, this can be accomplished frequently by shaking the door a little.

The rubber band method is effective and easy to use but leaves an undesirable mark in the door that will easily be seen on any well-kept door. Sophisticated intruders use a bent coat hanger rather than a tack to stretch the rubber band (Figure 6). The coat hanger must be bent properly and be long enough, so the door can be closed as far as possible. An even easier method, possible to use on doors with large enough space between the door and the jamb, is to insert a thin wire to move the chain back.

Sometimes the door will not be locked properly because

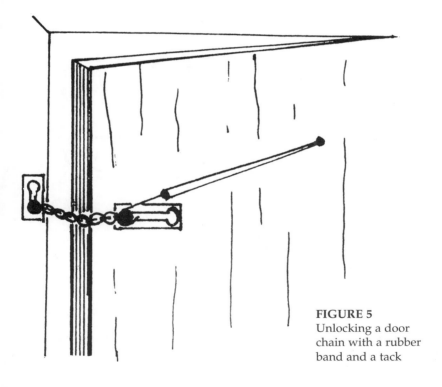

FIGURE 5
Unlocking a door chain with a rubber band and a tack

HOW THE BURGLER BREAKS INTO YOUR HOME

of a faulty door check, or door closer. A door check consists of a heavy spring and arm coupled to an air or oil cylinder that automatically closes the door. The door check also controls the speed in which the door closes. If the door check is not working properly, there is a good chance that the door will remain unlocked.

Destroying the door is not the only means of entering through a locked door. There are various other methods, most of which have been in use for quite some time, such as entering through a transom or manipulating the lock construction without actually picking the lock.

If a door warbles slightly and the lock is of an older construction that lacks a dead-locking function, the bolt can be retracted with a celluloid strip by the process known as loiding. This process involves slipping a flat object between the bolt and the strike. The strike, or striking plate, is the part of

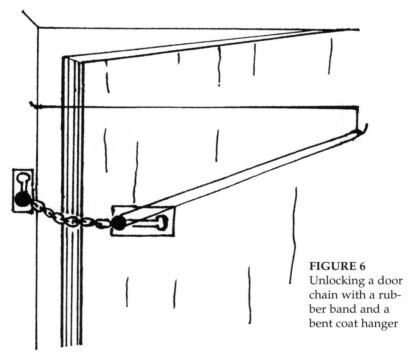

FIGURE 6
Unlocking a door chain with a rubber band and a bent coat hanger

the locking arrangement that receives the bolt, latch, or fastener. It is recessed in the door frame. A little pressure might produce enough space to insert a celluloid strip between the lock and the striking plate. Then only a slight pressure on the inserted celluloid strip will force the latch back and release the lock, thus opening the door. To counter this, you may consider fitting the door with a so-called antipick latch. This is a spring latch fitted with a parallel bar that is depressed by the strike when the door is closed. The depressed bar prevents the latch from responding to external pressure of any kind. This naturally makes loiding impossible.

A variation of this method is to insert a thin knife (e.g., a linoleum knife) between the door and the jamb. The burglar tips the point of the linoleum knife upward. A pry bar is then inserted above the linoleum knife to spread the door slightly. Then the latch can be disengaged. The burglar brings the linoleum knife forward, pushes the latch bolt back into the locking assembly, and opens the door. If there is enough space between the door and the frame, the linoleum knife alone is sufficient to move the bolt back. If this is the case, you should seriously consider changing either the door or frame or both.

Unfortunately, burglars can make enough space to insert a pry bar even where none exists by using wooden wedges. One wedge is inserted on each side of the bolt, about 10 to 15 centimeters away from the bolt assembly. This spreads the door away from the jamb sufficiently to insert the pry bar.

Modern doors and jambs sometimes fit so well that it is impossible even to insert a wedge. In this case, the intruder may use a stainless-steel shim. He will force the shim into the narrow crevice between the door and its frame and attempt to push the bolt back.

Another way of opening a door is to use a so-called Z-wire. This is a tool made from a stiff, thick wire, 25 to 30 centimeters long (Figure 7). The intruder inserts the Z-wire between the door and the jamb until the short end has reached all the way in and then rotate it toward himself at the top.

This will cause the opposite end to rotate between the door and the jamb, contacting and retracting the bolt. When the bolt binds, the burglar exerts pressure on the knob to force the door open.

If a transom has been left open or unlocked, there are a number of ways to enter. If the transom is completely open, any burglar can simply crawl through by stepping on the door knob if necessary. If the transom is only partly open, however, it might be impossible to crawl through. Then the burglar will lower a length of cord through the transom to form a loop that, when wrapped around and drawn taut around the inside doorknob, might twist it enough to open the lock, when he draws up on one of the two ends (Figure 8).

It is often easier to use two long pieces of string connected by a strip of rubber inner tubing or an electric cable covered by a strip of rubber than to use ordinary cord. The tubing will be 20 to 25 centimeters long. It is possible to open both a regular door knob and an auxiliary latch unit with this

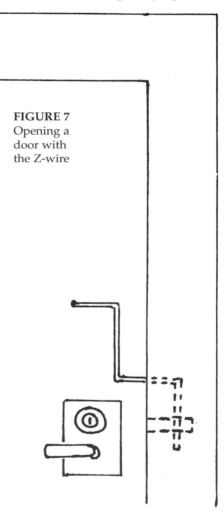

FIGURE 7
Opening a door with the Z-wire

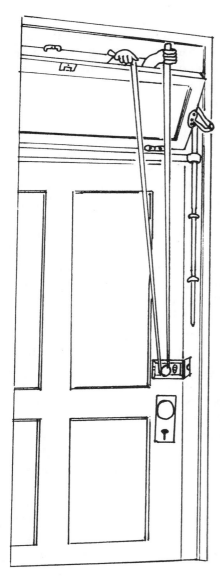

FIGURE 8
Transom entry

method, as long as a key is not required also from the inside.

Yet another means of covertly breaking in is to gently pry loose the molding around one of the panels of the door with a thin, flat chisel. If the panel can be removed without damaging it, the intruder can easily either crawl through the opening or at least reach through it and release the lock in the door.

In certain, especially tropical, countries, doorsteps are not very common. If this is the case, never lock your front door from the inside and leave the key in the lock. An old, but still not forgotten, trick is to insert a sheet of paper or an old newspaper under the door. The intruder will then push the key out of the keyway with a wire. As the key falls down and lands on the paper, he can easily pull both paper and key back and then use the key to open the door.

Windows

Windows always present special problems for the burglar, but typically he has techniques at his disposal to handle them. There are numerous types of window locks, for casement windows, sash windows, and most other types of windows. These locks can generally be picked from the inside, because they are extremely simple and sometimes all even use the same universal splined key. From the outside, however, the only means of entry is to break the window or lock open. And this is something many burglars are loath to do, because of the risk of hurting themselves on the splintered glass.

Unfortunately, the very construction of a window makes it relatively weak and easy to break. A forced entry through the window is almost always easy if the intruder cares to try it.

If the window is only closed and not locked, the easiest way to gain entry is to simply smash a small pane to allow one's hand through to release the catch. The window can then be opened. To prevent the window from being opened in this way, you need to install window locks. Even if the window is smashed, the frame will remain closed. To climb in through this frame would necessitate both the noisy breaking of a large part of glass and a quite dangerous entry, in which the intruder is likely to cut himself.

Of course, *you* cannot open a locked window without the key if you suddenly need an escape route in case of fire. You should consider keeping the key to the window lock close to the window. Make certain, however, that the key is not obvious from the outside and that an intruder cannot reach the key if he smashes the window. Remember that he may have some tool to reach into the room.

Louver windows are notoriously easy to force, especially from a flat roof (such as on a garage), offering easy access to the window. The louver window can be simply levered out of its frame, unless it is very firmly glued with epoxy resin. In this way, whole strips of glass can be removed from their metal clips. Modern louver windows might have locking devices installed (Figure 9), and sometimes the louver blades are also of laminated safety glass. Still, a louver window is always a favorite entry point for burglars.

There are various ways of opening an unlocked window from the outside, depending on the type of window. The easiest one to open is the sash window. This window consists of two halves, the top and the bottom, that slide up and down. The latch is located between these two halves. One of the oldest tricks of the trade is to slip a knife, shim, or similar device up between the sashes (the upper and lower halves of the window) to move the window latch to the open position (Figure 10). Then the window can be opened easily. If the area between the window halves is too narrow for a knife, a narrow hole can be drilled at an angle through the wood molding to the base of the latch. A burglar can insert a stiff wire through the hole and push the latch back. Afterward he may, if he bothers, camouflage the hole with paint or dirt. Contemporary fasteners, however, cannot be

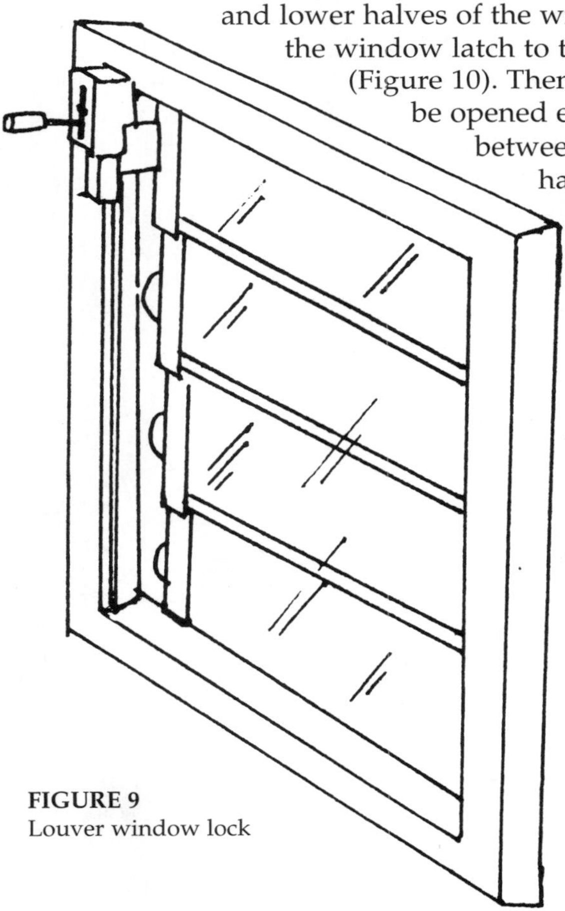

FIGURE 9
Louver window lock

HOW THE BURGLER BREAKS INTO YOUR HOME

manipulated in this way. For instance, the fitch fastener is a pivoting device with a snail-like cam that cannot be knocked back. Another type, the Brighton fastener, relies instead on a screw-down acorn that securely clamps the sashes together. These fasteners also sometimes contain integral locks. A lock is, of course, the safest security device in a window, and there is no reason at all why you should not lock all your windows, at least on the ground floor.

Sometimes the upper and lower window halves are connected by means of a bolt slipped into a hole drilled through the lower sash and partially into the upper sash. This was commonly done in many older buildings to conserve energy as well as to increase the security of the window. If so, an intruder can simply drill a small hole to the bolt and then push the bolt out of the upper sash by means of any pointed object (e.g., a nail) . The small hole will usually be difficult to see, and it, too, can be hidden with putty, paint, or dirt (Figure 11).

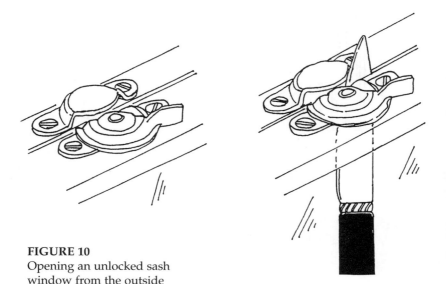

FIGURE 10
Opening an unlocked sash window from the outside

Other types of unlocked windows can also be opened from the outside if a narrow hole is drilled to allow a piece of wire to be inserted. These holes must be drilled in locations where they are much more likely to be noticed by the inhabitants of the house, but few burglars would care. All windows, including sash windows, can be fitted with locks. These come in various types, but all of them are of fairly simple construction. This will not assist the burglar greatly, however, because the locks are impossible to open from the outside except by first breaking the window. Even small padlocks may be enough to deter the burglar from making an attempt.

Metal window frames in older houses are generally made from steel or galvanized steel and, consequently, are quite difficult to break through. Modern aluminum-framed windows, however, are made of such a thin and soft aluminum alloy that the frames can easily be distorted with almost any lever. The self-tapping (self-threading) screws used in these

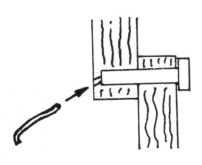
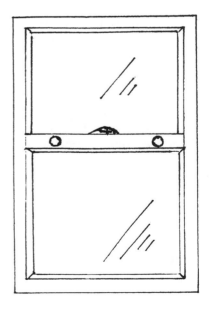

FIGURE 11
Opening a bolted sash window from the outside

constructions tend to pull out easily, and the locks, if any, are often of a very poor quality.

This is especially true of horizontal sliding windows. They work by sliding on an aluminum track. It is usually relatively easy to lift such a window out of its track, in the same way as lifting a sliding door, the process of which was described above.

Types of Glass

The important thing to consider when it comes to entering through a window is the type of glass employed in the window. Glass comes in several different types. The lowest quality is *sheet glass*. A cheap glass with imperfections, it is often referred to *as horticultural glass*, since it is often used for glazing greenhouses. It is generally not used for domestic glazing. Sheet thickness is 3 millimeters, and it breaks easily.

The standard material for domestic glazing is *float glass*. This glass is flat and free from imperfections. It generally comes in thicknesses from 3 millimeters to 10 millimeters. Four-millimeter-thick glass is very common, but the thickness is always related to the size of the window. A larger window needs glass of a higher thickness.

Wired glass is a rolled glass that has had a wire mesh embedded in it during manufacture. It is 6 millimeters thick and can come with either square- or diamond-pattern mesh. If the wired glass is hit by something, the wire will hold the glass together. Its resistance to impact is high, so it is often used where fire-resistance is important, as well as on glazed roofs where snow and ice are likely to drop on it. Wired glass is, however, generally not used for security glazing, because the mesh can be broken through once the glass has been smashed.

Laminated glass is true safety glass. It cracks but does not break under impact. This glass is formed by two sheets of float glass with a thin sheet of crystal-clear plastic sandwiched between them. It is very strong. Although the glass will crack under heavy impact, the plastic will hold it together very firmly. This is the most common type of security glass.

Tempered glass, also known as *toughened glass* or *armorplate glass*, is heat-toughened safety glass that is resistant to both impact and fire. When it does break, it shatters into numerous but harmless pieces with no sharp edges. Tempered glass is four to five times stronger that ordinary glass of the same thickness.

Many other types of glass, such as *patterned glass* or *solar control glass*, are also in common use but present no particular advantages or disadvantages to the burglar. They fulfill no special security purpose.

Of these various types of glass, only wired, laminated, or tempered glass will resist a sledge hammer, and the wired glass can still be broken through with the help of other tools. Most types of glass can eventually be broken, but it is sometimes both too time consuming and noisy to do so for the burglar to bother. If you are concerned about possible break-ins through your windows (e.g., in your shop) you may consider installing security glass.

Almost any burglar also wishes to avoid breaking through double glazing, because in effect this means that he has to break not one but two windowpanes. This doubles the risk of being overheard as well as of cutting himself. Of course, double glazing is also good in cold climates. Incidentally, double glazing will also make most outside bugs useless, because the glass no longer acts as the diaphragm in a microphone. This may interest those who fear electronic eavesdropping. While on this theme, it can be mentioned that if anyone fears being shot by an assassin armed with a rifle with infrared sights, an efficient but not very beautiful protective measure is to draw the curtains to prevent visual observation and then cover all windows with crated cardboard, which diffuses the heat signature.

The intruder breaking a window also dislikes venetian blinds because they create two different problems. First, it is often very difficult to see through the blinds to tell whether a room is unoccupied or not; second, to break through them makes a terrible noise. Burglars tend to avoid windows with venetian blinds.

SOPHISTICATED BURGLARY

This section is really only important to those who keep belongings in their home or office that are both known to possible burglars and also sufficiently valuable to attract the professionals among them. This means, among others, art collectors, gun dealers or collectors, and anybody who in his line of work keeps large quantities of cash or such valuables as gold or jewelry, which are easily disposed of.

If you are in one of these categories, your greatest risk is not the average opportunist burglar who just happens to pass by (although such can get it into their heads to attempt to rob you). Instead you need to worry about the professional burglar.

The Professional Burglar Knows What He Is Looking For

Professional burglars usually work in teams, either as part of organized crime or on commission (e.g., from crooked art dealers). This means that they known exactly what you have and what among your belongings they want. They will not waste time emptying your wallet left on the kitchen table or search your underwear for hidden gold trinkets, because they know all about your vastly more valuable gun collection in the basement. More to the point, because they know what they are looking for, they may not take kindly to accidentally finding you on the premises when they have work to do. While not necessarily more dangerous than the drug addicts who break into the home next door, more frequently they are armed and prepared to use violence. Unfortunately, there is an ongoing trend in organized crime to employ teams of professional burglars to break into almost any business with commodities that are easily disposed of. Paradoxically, the same diligent methods of break-in that were formerly reserved for goldsmiths' shops and the like are nowadays used to steal computer parts, TV sets, and even foodstuffs on their way to the supermarket.

The professional burglar will spend a comparatively long time on target reconnaissance and initial planning. Much of this can be done without breaking any laws or at least without leaving any evidence that can lead to an indictment. Although you cannot do much to counter him at this stage, you should know what a burglar is looking for. For a serious break-in aimed at a well-protected office, a number of data about the target is required. First, the intruder will attempt to determine the following factors about the target:

Protection
- Number of guards, if any.
- Location of guards at particular times.
- Their equipment and armament.
- Technical details on alarm and controlled-entry systems, if any.

Layout
- Number and locations of entrances and exits (both ordinary and emergency exits), and hallways or stairways suitable for a quick escape. Note that windows, rooftops, and sewers might also be used for this purpose. Even chutes might do, in an emergency.
- Location of important offices or other rooms of importance.
- Method of smuggling the team of burglars and their equipment into the building, if ordinary access is denied.
- Possibilities of temporarily cutting off the entire building from the telephone network and other ways of preventing communication with the outside, for instance with cellular telephones or radio transmitters.

Personnel
- Number of staff members in the building during and after ordinary office hours.
- Their location at particular times.

Timing
- When the target is most vulnerable.
- Whether there are any outside factors that will influence the accomplishment of the break-in.

Targets are typically most vulnerable in the early morning between 2:00 and 4:00, since most people are asleep at that time and any security guards tend to be sleepy and less vigilant during these hours. Different individuals obviously keep different habits, so the intruder may wish to determine the local conditions.

One way of getting information on the tenants of a particular building is to play the role of a private detective. Such detectives are sometimes prying into the comings and goings of tenants, especially in divorce cases. Another way of obtaining information on tenants is to go through their trash. An office especially yields a surprisingly large amount of trash, most of it in the form of discarded documents. This can give much valuable information. To some, this method is even useful in determining whether a break-in is really called for or not.

Torn or burned scraps of letters might be found that indicate whether the break-in will be fruitful or not. (Burned or dirty pieces of paper can be read through the use of special equipment.) It is not unheard of that, although a businessman very carefully destroyed his important letters, his secretary simply discarded her stenographic notebooks in the wastepaper basket, where it was found by a delighted sleuth who was trying to piece together the businessman's activities.

Another obstacle might be an alarm system. And as we learned earlier, alarm systems that frequently sound the alarm because of technical defects will not be taken seriously by investigating police patrols or private security companies. A burglar may, therefore, over a period of several days or weeks, at times discreetly alert the alarm system and quickly withdraw from the area without leaving any traces. Then, during the actual break-in, the police will be much less vigilant. This allows the burglars a longer time to do what they came for.

This might also be true of an ordinary guard. During the early years of World War II, when he worked for the Office of Naval Intelligence, Willis George planned and executed several break-ins. During one of these operations, he noticed that a foreign consul had posted an armed guard in the private part of the consular office. Obviously this made a covert search impossible. So George had to devise a means to get rid of the guard without exposing his own operation. The guard had probably been posted there, because the consulate elevator operator had become suspicious during a previous covert search in the building.

George finally decided to try to make the elevator operator appear ridiculous by behaving in an overzealous manner. If the consul decided that the guard was unnecessary, then he would likely dispense with such extreme measures for protecting his office.

For this purpose, a few nights later George again entered the building and deliberately made some noise to alert the elevator operator. Then George hurriedly left the scene. About half an hour later, the consul arrived by taxi. Just as was expected, the elevator operator had called for him. They searched the office, without finding any signs of a break-in, of course. The consul was angry when he left after the fruitless search.

A few nights later, George repeated what he had done. Upon hearing the suspicious noise, the unfortunate elevator operator immediately called the consul. The consul arrived, angry and tired from being awakened every other night. Again their search produced no indications at all that a break-in had been attempted. The next night, the guard was no longer posted in the office, and George and his team could continue to prepare for the actual break-in.

In this context, it should also be noted that the majority of all false alarms take place in the morning between 7:00 and 9:00 and in the evening between 5:00 and 7:00. These are the times when the ordinary occupants of the buildings protected by them turn these system off or on. The police and serious security companies know this and therefore are less vigilant during these hours.

Outside factors that might affect the break-in can be of any type. For instance, in a government building or an industrial complex, especially one of military importance, consider the possibility of a surprise inspection by a senior official and a large number of guards. This is perhaps a worst-case scenario for the burglars, but other things may also affect the break-in, such as the sudden removal or installation of important machines. Such work, and also regular maintenance work, is often performed outside ordinary office hours so as not to delay the regular work. Sophisticated burglars know this and prepare for such eventualities.

The Professional at Work

For the rest of this chapter, we will take the unusual step of detailing how the elite of the world's burglars plan their break-ins. No longer are we dealing with the day-to-day crime that affects most people or even the fairly sophisticated burglar who comes for your private gun collection. No, this is how the professionals who go for bank vaults and valuable art treasures do it. Read and take warning if you consider yourself a possible target of such men.

In addition to the steps described above, the intruder also carefully reconnoiters the area around the target. Suitable escape routes will be prepared, both for use after a successful break-in and for use in the case of failure. If there are a number of burglars involved in the job, they may organize an outer security ring around the target during the actual break-in. The members of the outer security ring are positioned some distance away from the actual target. They observe access routes and warn of the approach of police, either on foot or in vehicles. Warnings are generally passed on by radio, preferably by the use of innocent-sounding code words, so that the transmission cannot be identified if it is monitored. For more information on this procedure, see my book *Techniques of Secret Warfare: The Complete Manual of Undercover Operations* (Miami: J. Flores Publications, 1996).

At this stage, every burglar on the team may feel the need to familiarize himself with the appearance of everybody working in the building at the proposed time of entry and of every resident or employee of the target apartment or office. In this way, the members of the outer security ring can easily recognize any potential threat and sound the alarm in time to warn the burglars if somebody is approaching the target area.

This familiarization process has often been handled by sending the members of a team to the building in daytime dressed as cleaners, maintenance personnel, or repairmen. They can then spend plenty of time fixing some minor problem or paint a wall or another object and at the same time get a close look at the entire staff of the building. One good way of identifying everybody is to litter the floor with tools and then warn passersby not to stumble on them. Of course, a sufficient cover is required for this kind of job, including the address and telephone number of the company that is supposed to have sent the repairmen there and the name of somebody on the premises who has called for them in the first place. The name of the company will also be prominently displayed on both their car and the boxes or bags containing tools the repairmen bring into the building.

The normal amount of traffic in the building at the proposed hour of entry will also be assessed carefully. The behavior and patrol pattern, if any, of the watchman or the security guards will be noted particularly and compensated for in the final plan.

The reconnaissance phase is followed by the planning

phase. First, it must be decided whether the entry is to be covert or overt.

A covert break-in is generally to be preferred because, if successful, this will give the burglars far more time to escape and dispose of the stolen goods. Any subsequent police investigation will also have the disadvantage of following a cold track, with all that this implies with regard to forgetful witnesses, among other factors. However, a covert break-in is difficult—sometimes impossible—for the burglars to execute and has a much greater chance of failure.

As a curiosity, it may also be mentioned that the covert break-in is also a procedure occasionally used by the police to acquire information without having the proper search authorization. Of course, such an operation is illegal in most jurisdictions.

In either case, the planning phase consists of more or less the same type of work. It must be realized, however, that the planning described in this chapter might take several weeks to complete. Consequently, a break-in of this complexity will not be undertaken on the spur of the moment.

First, if the target is located in a block of flats, the burglars must decide how to enter the building that contains the target office or apartment. This is frequently done by renting another flat in the same building. Sometimes it is possible instead to enlist the help of the owner or the superintendent or some other worker or resident in the building. The superintendent is preferable, because he frequently has a master key to all apartments in his building. In either case, the personal characteristics and loyalty of the helper are researched thoroughly to preclude his informing the police at a later stage. If the man turns out to be untrustworthy or none of these alternatives is available, the burglars must find a plan that encompasses both entering the building and executing the subsequent break-in.

The procedure is more or less the same whether the target is a villa, small house, or mansion. A neighbor might be available, or the team of burglars must enter the neighborhood and perform the actual break-in at the same time.

Securing a master key is sometimes possible in flats and office complexes because the caretaker generally needs one. The same is true of hotels. It is generally impossible in villas and mansions owned by private individuals. Despite this, it might be possible to obtain and make a copy of the relevant key prior to the break-in. A way of doing this is by searching the pockets of the owner of the key when he is in a situation in which he is unable to notice the search. Such opportunities are not easy to find, but might occur in a locker room or gym.

A key is not only useful when entering the premises, it is also invaluable if the burglars have to make a quick escape.

As noted above, a sufficient number of escape routes will be prepared. If a master key, or the actual key to the target area, can be secured, a sufficient number of copies will also be made to ensure that no member of the team is trapped in the building in case of an emergency. This surely is a vital precaution: the burglars know that any associate of theirs caught by security or the police is likely to offer testimony to save his own skin.

The members of the security ring will also aim to actively delay any employee or resident of the target apartment who might want to enter the premises when the burglars are working there. Such a delay can be caused by a member of the security ring pretending to be drunk and accosting the employee until the intruders have managed to withdraw or perhaps pretending to be involved in maintenance of the elevator. The use of a minor car accidents or any other plausible delay is also possible. Here creativity is valuable, because not even the most skillful burglar can plan for every possible situation. For this purpose, an inner security ring is sometimes organized to deal only with such situations.

It is not always necessary to leave the building even if the security ring gives warning. Sometimes the team of burglars can simply retreat temporarily to a nearby office or apartment and wait for the moment when they can resume their work. This option is generally not available during overt break-ins, however.

Every member of the team may even be supplied with a convincing and documented cover story explaining what he is doing on or near the premises. If so, the contents of his pockets and his clothes will conform to the cover. The equipment brought onto the premises, as well as all bags, can also be expected to conform to the cover.

Some burglars may even rehearse the break-in several times in a safe location so that every team member knows exactly what he is to do and how he will do it. The actual break-in will be performed if possible without talking or even discussing what to do, as long as the team is in the target area.

How Your Government Does It

The real break-in specialists are found in certain government agencies. They specialize in covert break-ins, usually as part of espionage rather than law enforcement operations.

A covert break-in is much more difficult to perform than an overt break-in. Speed is important in this break-in, too, because every minute in the area brings a chance of discovery and capture. But even more important are skill and diligence. Locks are picked, and the means of entry, if successful, never leaves any marks revealing that a break-in has occurred. Actually every part of the premises is to be left in exactly the same condition it was before the break-in. This requires considerable skill. Furthermore, a covert break-in must frequently be called off because a certain lock or safe proves impossible to pick in the time available.

When all the details of the planning phase have been taken care of, a preliminary break-in is sometimes undertaken. This break-in is very cautiously executed, and no attempts are made to find or obtain any interesting documents or other evidence in the target area. Instead, the break-in is only to make certain that the plan is valid and that sufficient expertise is available to perform the actual break-in when the team and its equipment will be too conspicuous to allow for failure.

The preliminary break-in is made in absolute silence and with extreme caution. The need for silence is because voice-activated tape recorders might be hidden in the target area. There might also be any number of innocent-looking traps, such as a short length of thread or hair, a paper clip, or small objects positioned in a certain spot, or a special arrangement of some books or papers. All these traps alert a cautious target that somebody has been tampering with his things or had entered the premises. Therefore, before any object is moved, the team takes accurate measurements of the position of every object, usually with the help of a Polaroid camera.

There might also be more devious traps, such as a video camera and recorder initiated by any type of sensor capable of detecting an individual on the premises.

Many computers register the time they're turned on or execute a command. It is very risky to check the contents of a computer, unless the burglar knows exactly what he is doing. Of course, this is never a problem with forced break-ins: computers are valuable objects, and so the entire set can easily be taken, without anybody's wondering why the burglars stole it.

Yet another problem is taking objects from areas covered in dust. The burglar must not remove an object from such an area unless he can replace the dust in a convincing way. A small atomizer filled with talcum powder mixed with powdered charcoal is sometimes used to simulate dust.

During the preliminary break-in, the burglar also makes a detailed check of the premises. For example, during the actual break-in, there might be a need for blackout curtains. The burglar must then determine exactly how many such curtains (as well as their required size and the method of affixing them to the windows in question) are required. He also considers at this time the number and types of safes and file cabinets. As many details as possible about them, as well as about all ordinary locks in doors, for instance, are collected, including any numbers on the locks. These details are frequently necessary for the locksmiths involved to know in advance to pick the locks during the actual break-in.

Government-sponsored burglars also need to find a safe place for positioning the camera equipment and Xerox machine used to copy any found documents during the actual break-in. Such a place is typically not located in the office or apartment to be searched, because this might preclude or hinder a quick escape. A small nearby cleaning equipment storage room is ideal for this purpose. Toilets should be avoided because they might be visited by someone unexpectedly going to his office or apartment during the break-in. If the chosen camera site is sufficiently remote, it might even be possible to return later to recover the equipment hidden there, in the event of a quick escape becoming necessary.

It is apparent from all this that a covert break-in requires very careful planning, as well as specialized skills to perform it successfully. Besides this, it is necessary to maintain total security against interruption during the search of the target building. This necessitates the use of an outer security ring, and probably an inner security team as well, able to delay any passersby.

A covert burglary team usually consists of several individuals, each of them an expert in his field. Among them are a lock-picking expert, a safe expert, and one or more experts on alarm systems to execute the actual break-in. The alarm systems experts are also prepared to manipulate electronic locks and other electronic systems (e.g., elevator machinery).

Furthermore, government-sponsored burglars include one or more analysts capable of rapidly evaluating any found documents and at least one photographer. The photographer carries several cameras, both for copying any found documents and, equally important, for snapping Polaroid photographs of the original appearance of the rooms to be searched, thus providing a pattern for restoring everything to its original position before leaving the premises.

As noted above, it is also common nowadays to use a

small, portable copying machine to duplicate any found documents. Of course, an adequate supply of film, paper, and so on, is included in the equipment brought onto the premises. Sometimes even infrared photography techniques are used, because they require no visible light source.

Government-sponsored burglars also include an expert in opening letters, capable of opening and resealing any type of letter in a convincing way. This technique also requires a certain amount of equipment, maybe even an infrared fluorescence detection system to detect traps on sealed letters or documents or alterations to passports or other documents.

Finally, any guns or other means of dealing with individuals interrupting or noticing the break-in will be decided on. Burglars of this sophistication can be assumed to be both armed and ruthless. They have too much to lose.

Every piece of equipment, including radio transmitters, is first tested and checked, so that nothing is found missing after the team has entered the premises. Radio transmitters, especially, might require a test in the area of operations before the actual break-in is initiated. The range of radio transmitters depends very much on the construction of the building; different buildings disturb radio transmission in different ways.

A spare camera might also be brought; a broken camera in the middle of a break-in is a bad reason to pack up and go home.

During the actual break-in, the premises may initially be entered by only one burglar to ensure that the entire group does not walk into a trap. When this pointman gives the agreed-upon signal, the rest of the group will follow.

When on the premises during a break-in, sophisticated burglars avoid the use of any toilets or water taps. This might leave traces or, more important, make noise. They also avoid smoking, since this leaves both traces and odor.

Finally, when leaving the premises after a covert break-in, whether it was successful or not, prudent burglars check to see that no piece of equipment has been left behind and that the appearance of the target building is exactly the same as before the break-in. This might require polishing or even rewaxing the floor if the burglars have entered with their shoes on. Of course, all fingerprints are wiped away, whether on doors, walls, or office equipment. If a safe has been opened, the dial is reset to its original reading. If the flat contains thick rugs on the floor, it might even be necessary to sweep them upon leaving, so as not to leave any footprints.

We have described the methods and tactics of burglars far more dangerous and resourceful than you are ever likely to meet—unless your line of work is very special indeed. So let us now return to the day-to-day crime that most readers of this book are likely to be affected by.

COMMON MISTAKES THAT GIVE THE BURGLAR AN EASY RIDE

Ordinary burglars use numerous tricks of the trade to find information about or gain admission to a home or office. These tricks include ways to find plans of the target's alarm system and means to recognize traps and faked obstacles to the break-in. Common hiding places of keys and combination codes also fall under this heading. Not all can be detailed here, but at least some of the most common tricks should be mentioned. Be on your guard against them.

The easiest way to enter the residence of an unsuspecting owner is to use the real keys. If the owner leaves the keys outside his house when he goes away, this is quite easy. Any burglar who knows his business will check the seven most common hiding places for keys:

- Under the doormat
- Under a flowerpot
- In a flowerpot
- Under a stone alongside the path or near the front door
- Stuck or hanging under the window ledge

- Inside the door on a piece of string that can be pulled through the mail slot in the front door
- Just inside an unlocked garage or storage shed

If the intruder can gain entrance to the house for a few moments on one pretext or another, he might be able to check whether the key is left in the lock on the inside or in an obvious cupboard or hook near the front door. If so, he might be able to gain an unsupervised moment to make a pattern of the key in a wax box. Then a copy can be produced at his leisure.

Sometimes supposedly locked doors are nevertheless found unlocked or even open. The reason is usually very simple: to facilitate the work going on inside or the delayed arrival of a visitor or employee. Some people regularly leave doors unlocked, especially in offices, so they don't have to use a key when they return. But the human factor is not always the reason for an unlocked door. The lock may be improperly installed, so that the door will not lock even when closed, or the door check is out of order.

For this reason, many companies keep a guard or a caretaker who regularly checks that all doors are locked and remain so. If the guard follows a fixed routine, however, any reasonably competent burglar can plan the break-in so that the guard is safely out of the way. Otherwise a burglar could position a partner to watch for the guard to know when the guard is doing his rounds.

The points of entry include, but are not limited to, doors, windows, and any other openings that provide access to the interior of the target building. As mentioned above, a burglar sometimes finds it easier to enter through a roof or ceiling.

Most houses protected by an alarm system have a sensor on each outside door or window accessible from the ground and large enough to crawl through. Few homeowners protect the windows on the second floor in this way, even though the windows may be easily accessible from a flat garage roof or nearby tree. You should not be one of these negligent homeowners, especially if the windows cannot be seen from the street or a neighboring house. The burglar will try to pinpoint the sensors before the actual break-in attempt: do not offer him any opening that you have overlooked.

As every burglar knows, there are certain indications of whether a house is empty or not. A mailbox stuffed with old letters, circulars, or advertisements shows that nobody is at home. Another indication, maybe the most important one, is the absence or presence of lights.

Many houses today are equipped with outside lights that are switched on at dusk every night by a light-sensitive photoelectric cell controller, which measures the intensity of natural light. It is triggered by the natural light levels and switch on the light at the right moment, just when the light falls below a predetermined level. The lights are switched off automatically when daylight returns.

Other buildings rely on programmable automatic timer switches that turn on the lights at a predetermined time every time. Some of these timer switches are of the 24-hour variety that will turn the lights on and off at the same time every day. Others are of the seven-day type, that can be set to allow the lights to come on and off at different times each day of the week. The switching pattern is repeated only after a week has passed. Sometimes the users of these devices do realize that they must vary the setting of the timer with the changing of the seasons. You should be one of them. There are devices available for this very purpose. They can be programmed to vary the on and off times that the light, or lights, will be turned on each day. These devices are called solar dial timer switches and are programmed for the seasonal changes in the day's length.

Such devices are often both powered by and transmit their signals over the electric wiring in the building, although remotely controlled units using radio waves are also available. The lighting may even be connected to the output from the alarm system, so that all, or at least most, lamps in the building will turn on or start to flash whenever the alarm is sounded. Other appliances, such as a stereo system, can also be activated in this way (do not link a timer to a television set, because this creates a fire risk). Virtually nothing is impossible, but the sobering fact remains that this will not prevent an experienced intruder from breaking in.

Finally, passive infrared detectors might be used for turning on the light as soon as somebody approaches. Such a detector is easily wired to a light source, and the light will remain on as long as there is somebody in the vicinity emitting body heat or even for a set period after the source of warmth has gone. Such an infrared floodlight system also includes a built-in photoelectric cell that prevents it from being triggered in daytime. Sound-triggered detectors are also available for the same purpose. Since burglars do not wish to work in bright light and full view of any passersby on the street, such lights are often effective deterrents. But do not position the light so that it shines into the windows—this will prevent you from seeing into the garden. Also make sure that the light does not cast shadows near the windows, thus providing a burglar with cover to open or attack them, or near the door, giving the burglar a place to lie in wait to attack you.

Another security device is the electric curtain controller that opens and closes corded curtain sets. This device is generally connected to a timer for automatic control.

A radio, preferably tuned to a talk station, can be left turned on or connected to a timer switch. Finally, a telephone-answering machine programmed with the message that the owner "cannot get to the phone right now" (not "I'm not here" because this will invite burglars by informing them that nobody is at home) is also frequently used to leave some doubt whether anybody is at home or not.

By all means use such devices to pretend that you are at home, up, and awake. However, these means of scaring away ordinary burglars are not very effective. It is generally easy to tell whether anybody is at home or not by calling and pretending to be a salesman. This is a classic pretext among burglars, and it never fails. Furthermore, some people leave the light on only in the hallway rather than in the living room. This is a definite indication that the house is empty—nobody actually *lives* in the hall. Leaving the light on in a downstairs room with the curtains closed is slightly more clever but not enough to deter a professional intruder. It is slightly better to leave the light on in a upstairs bedroom, where an intruder cannot peer in to check whether it is occupied or not. Even so, an experienced burglar will not be fooled.

It should be mentioned that by cutting the electrical power or telephone lines, a burglar can determine effectively whether anybody is at home or not: furthermore, most of the previously mentioned protective light and alarm systems are rendered useless, and the owner is certain to reveal himself if he really is at home. The burglar knows this. If there is no reaction, he will go in.

Some people use recordings of a barking dog that are activated whenever the doorbell is rung. There is even a self-contained alarm system available that electronically imitates the sound of a dog whenever somebody approaches the protected area and is detected by a passive infrared detector. These various dog-imitation systems can always be identified by the fact that you will only hear the actual barking and no other noise, such as from the dog running around or jumping at the door.

Real dogs are a very serious problem for the burglar, especially if they bark at strangers and generally appear unfriendly. Alarm dogs are not always dangerous; many are kept mainly as a psychological deterrent to frighten away intruders rather than actually attack them. Terriers make excellent alarm dogs for this very reason. Still, the barking of a dog calls unwanted attention to the scene, so the average burglar will avoid dogs or at least attempt to silence them in some way.

Certain large dogs, such as the Great Dane, mastiff,

Alsatian, Doberman pinscher, and rottweiler, are sometimes trained to not only raise the alarm but to defend the home against attack. Really dangerous patrol dogs are not usually kept as domestic watchdogs, however, because they are simply too dangerous unless in the care of a professional dog handler. No burglar wants to come across such dogs.

As a curiosity, it might be noted that in rural areas geese are also a very real obstacle. Geese are highly territorial and not only honk furiously but often chase and peck any intruder who does not retreat quickly enough.

Many blocks of flats employ an audio entry system. This is a means of access control in which a speaker panel is located outside the premises, near the front door. The speaker is linked to a telephone handset or microphone device that permits two-way communication. The device is frequently used in conjunction with a digital code lock entry system. Each flat in the building is then equipped with a telephone handset as well as a remote control that releases the electrically operated door lock.

The audio entry system might also include an audio-visual component. It will then function as the standard audio entry system but with the addition of closed-circuit television.

It is, however, all too easy to circumvent an audio entry system, the key being the human factor. The burglar will just call up anybody in the building and either say that he has lost the code but a "friend" (who he knows is a resident in the building and whose name he will mention) asked him to go inside and take care of something in his flat. If this does not work, he will call up another resident and say that he has to enter to deliver something (not mail, though, because the postal carrier has the correct code and does not need to ask for it). As a final method if the others fail, he can always say that he wants to come inside to deliver free samples of something that he believes is desirable to all households in the block. Of course, the burglar might also wait for a resident or somebody with the code to arrive and then hurry in with him or her.

There are certain preferred ways of gaining entrance to a private home during the preliminary reconnaissance of the premises. These methods are also frequently used by con-men. The intruder dresses in a certain way (e.g., as a serviceman for one of the public utilities companies) and approaches the target. For example, claiming that he needs to check the homeowner's electric power, the "serviceman" asks the owner of the house to stand by the fuse box and turn the power on and off while he goes around the house to "check the circuit." If he is posing as a representative of the gas company, he might ask the owner to watch the meter outside, while he "checks the appliances." A "water company employee" might ask the owner to turn the taps in the kitchen, while he "checks the flow" in the upstairs bathrooms, basement laundry room, or other part of the house. Someone posing as an employee of the tax assessor's office or the census might ask to measure or count the rooms for assessment or accounting purposes.

Burglars trying any of these impersonations do run the risk that the owner of the house will ask for official identification and then lock himself into his house or flat while calling the local office to inquire about the intruder. Some burglars attempt to foil this by printing a friend's telephone number on the fake identity card. *Homeowners: Do not be taken in by this simple ruse.* The real government office number is listed in the telephone directory.

Some burglars attempt to get inside by asking to use the telephone for an emergency or even asking for a glass of water. It is amazing how easily most individuals are taken in by such simple tricks, especially the elderly. Burglars often work in pairs, with one engaging the resident in conversation while the other ransacks the house for valuables. For the careful intruder it is always worth a try, as long as an escape route is prepared in advance.

A burglar might pass himself off as a salesman. Then he might offer a free estimate for wall-to-wall carpets or new kitchen cabinets, or anything else that will justify his entering the home and proffering something very attractive in price. A variant of this theme, in rural areas usually, is the itinerant antiques buyer who wants to browse around the house looking for interesting pieces to purchase. Also in rural areas, or where wood or coal fires are common, a burglar can disguise himself as a chimney sweep. A common trick to be left alone in the house is to tell the owner to go out into the garden and shout when he sees the brush coming out of the chimney.

Burglars like keys. Do not keep your keys marked with your name and address or in a wallet together with other information of the same kind. If you drop your keys or wallet, any finder will be able to get into your home. A burglar may even call you, pretending to be with the local police, to inform you that your keys have been found and can be picked up at the police station. While you rush there, the burglar—who has been observing you—will enter your home.

Likewise, if your misplaced keys are ever returned to you, the only safe thing to do is to have the locks changed. In case of cylinder locks, you can change the cylinder, which is less expensive than replacing the entire lock. Of course, you also need to change the locks when you move into a new home.

It is a great advantage for the burglar if he knows in advance of the existence and type of alarm or locking devices in the building. In some countries, notably the United States, many cities and counties require a permit for installing an alarm system. This means that the relevant archive might be a good source of information about the possible existence of such means of protection.

The plans for an alarm system are often kept in the office of the security company that originally designed it, along with the installation and operator codes. If a burglar can gain entrance to this company's office, he can take advantage of this information.

Finally, it must be remembered that for commercial buildings the local fire department also have a say when a major alarm and security system is designed. For instance, this might mean that the fire station has a copy of the plans, including master keys in certain cases. It might also mean that some exits will be left open, or almost open, on purpose, regardless of the risk that an intruder could enter there: the safety of the personnel working on the premises generally has a higher priority than keeping intruders out. This is especially important in a building where burglars are not expected to be interested in the merchandise or machinery.

FILE CABINETS AND DESK DRAWERS

This is a good place to include a short discussion on file cabinets and desk drawers and which tactics and techniques burglars use to open them.

When a burglar has already entered your home, any locked file cabinets and desk drawers will not slow him down. He will not hesitate to force them open, usually destroying the cabinets and drawers in the process. I have even seen burglars slash locked briefcases with a knife to see what is hidden inside. At this stage, any additional noise will not rouse the neighbors, and a locked drawer is a sure sign that its owner is hiding something of value there. The moral is, never bother to lock any file cabinets or drawers at home, unless you wish to hide their contents from the members of your family and do not really care if a burglar damages them (and steals the contents).

You may think that you need to keep file cabinets and drawers locked so that your employees cannot get at the contents. Think again! There are various ways to open file cabinets and desk drawers, determined by the construction of the cabinet or drawer and the type of lock used in the object. However, with the exception of most sliding cabinet doors, the design is more or less standardized. This means that if you have really sensitive stuff, keep it in a hidden safe. Otherwise do not bother. Any locked drawers will only attract unwelcome interest, be it from burglars, your children, or employees.

If you think otherwise, see how easily these locks can be opened even without smashing the object. For example, a file cabinet. The locking device of file cabinets relies upon a bolt that extends from the top of the lock body. The bolt manipulates the control bars that lock each drawer of the cabinet in place. Because the bolt is spring-driven, it will extend out of the lock at all times, unless the lock plug is turned. The plug has a cam attached to the back, which rotates in conjunction with the plug. This cam works in a notch on the side of the bolt, thus operating the bolt (Figure 12). The cam can also be the actual lock bolt, especially in many drawer locks.

Locks of this type are generally not very secure. They also usually have open keyways, the keyway running from the face all the way through the body of the plug. This means that the lock can be opened most easily by jimmying. A jimmy is a pointed tool made of a thin strip of spring steel 3 to 5 millimeters wide. An intruder slips the jimmy through the keyway and manipulates the bolt directly, pulling the bolt downward, without bothering to actually pick the lock (Figure 13).

Some contemporary locks of this type do contain some kind of protection against jimmying the lock open. For instance, the keyway might be blocked by a piece of metal or a pin so that no access is allowed to the bolt. Although an intruder can pick such a lock, an alternative method is to use a rod of stiff wire not more than 5 or 6 millimeters in diameter and at least 30 centimeters long, with one end turned 90 degrees. This rod can be inserted between the drawer and the cabinet face. The bolt mechanism can then be forced down with this wire (Figure 14). In most cases, the rod is turned counterclockwise.

Yet another method also involves prying the drawer open with a thin piece of string steel or a wedge far enough to allow direct manipulation of the bolt. A strip of steel, about 45 centimeters long, between 1 and 2 centimeters wide and half a millimeter thick (or an ordinary letter opener can be used), is then inserted between the drawer catch and the

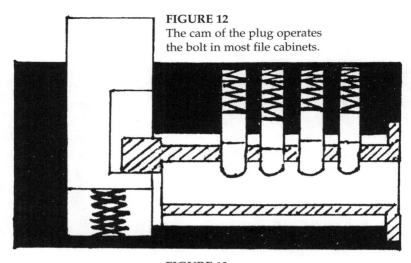

FIGURE 12
The cam of the plug operates the bolt in most file cabinets.

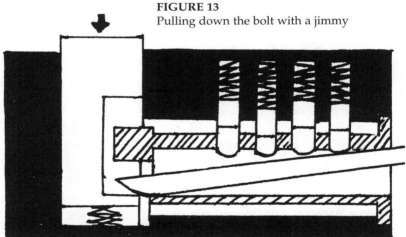

FIGURE 13
Pulling down the bolt with a jimmy

bolt mechanism to pull the drawer open. The opening tool will create a bridge for the drawer catch of the mechanism to ride on and pass the bolt (Figure 15). Usually the intruder must pull quite hard.

Older types of file cabinets can be opened by simply

spreading the file cabinet from its frame with a screw driver or some other sort of prying tool, while a thin but strong tool (e.g., a hacksaw blade about 30 centimeters long) is inserted to lift the bolt mechanism away from the drawer catch. Amazingly, I have seen lockable compartments in high-security safes, no less, that could be opened in exactly this way in a matter of seconds. Why bother to lock?

Certain types of file cabinets function in the same way, but with the mechanism effectively out of reach, hidden in the rear portion of the cabinet. Then it is sometimes possible to tip the file and in this way gain access to the mechanism, forcing it upward to release the drawers. If the cabinet is designed in this way, the mechanism can be seen protruding through the bottom partition of the cabinet.

Yet another method that works with certain types of file cabinets is to tilt the cabinet backward about 15 to 20 centimeters off the floor. The intruder then suddenly releases the cabinet so that one drawer catch disengages itself from the catch on the mechanism rod. The drawer is then released. The remaining drawers can now be manipulated open by hand, simply by reaching inside the cabinet to release the rod that keeps them in the locked position.

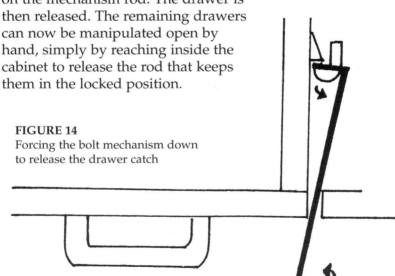

FIGURE 14
Forcing the bolt mechanism down
to release the drawer catch

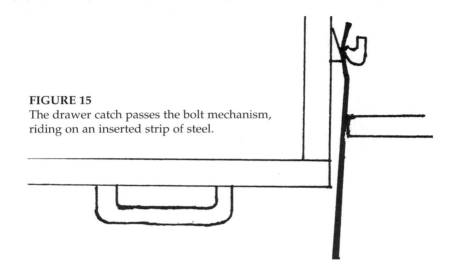

FIGURE 15
The drawer catch passes the bolt mechanism, riding on an inserted strip of steel.

Certain file cabinets have gravity-type vertical engaging bolts instead. They can be released most effectively by inverting the entire structure.

Although most file cabinets have the lock in the same position, there are various ways of arranging the locking bar and locks in a group of desk drawers. If all other methods fail, the serious intruder who does not wish to leave any traces after his break-in will pick the lock itself. He will, no doubt, first find a cabinet of identical type and practice on it before the actual break-in.

Desks with locking drawers controlled from the center drawer can be opened in another way. Many of these desks are constructed so that there is a space between the back panel of the desk and the back of the desk drawers. The locking bar is usually designed to engage the desk by either upward or downward pressure. If it is upward or downward depends on the style of bolt used in the design. The spring-loaded bolt is automatically pushed into the locked position by the motion made when the locking drawer is closed. To open the side drawers, the bolt needs to be pushed up by hand from under the desk, reaching up between the back panel and the back of the desk drawers. When the bolt is raised, the hook catches are released and the side drawers can be opened. The center drawer has its own lock, of course, which the burglar might have to pick.

Other types of desks might require the use of a little force coupled with pulling outward on the center drawer. This will push the bolting mechanism downward just enough to open the various drawers.

The center drawer lock can be opened in the following way. The intruder first inserts a screwdriver or other prying tool between the drawer and the underside of the desk. It might be necessary to protect the underside of the desk from scratching with some cardboard or thick tape. Then he pries the drawer away from the desk top, after which he uses a similar tool to pull the drawer outward to open it. This process might damage the drawer, so he may find it easier to smash the drawer in the first place, if so inclined.

A final way of opening the lock, but one that leaves clear traces of the opening attempt, is to drill a small hole in the drawer above the lock. The intruder inserts a stiff piece of wire (e.g., a paper clip) into the hole to push down the plug retainer ring. This pulls the plug free of the lock, which will cause the bolt to drop down into the open position. The resulting hole can be partly hidden by inserting a wooden plug of the right material, but it will be revealed by a careful investigation.

FIGURE 16
Sliding cabinet door plunger lock

A sliding cabinet door is often locked by a plunger lock (Figure 16). Such a lock is mounted on the outside door. The bolt is a projection from the rear of the plug unit that engages a hole in the other sliding door of the cabinet. Neither door can move since they are locked against each other. The plug is returned to the open position when the correct key is inserted into the lock and turned. The actual lock is either a disk tumbler lock (in old buildings) or, more commonly, a pin tumbler lock (for technical descriptions of these lock types, see Chapter 6). The easiest way to open these locks is to spread the two doors far enough apart to disengage the bolt. Alternatively, the lock can be smashed or picked in the ordinary way.

CHAPTER 2

How to Select a Reliable Alarm System

As we learned in Chapter 1, almost all alarm calls are false. Thus, a single alarm call may not be such a great threat to a burglary. However, the burglar may be worried by multiple alarm calls from the same object, because this increases the chance that it is a professional system and that security guards will actually come to investigate.

THE SECURITY SYSTEM'S THREE LINES OF DEFENSE

A comprehensive security system consists of separate subsystems set up in several lines of defense. Not every homeowner needs to consider all these subsystems. We will go into the details of these various defenses later, but here is a brief overview.

A reliable security system is typically divided into three lines of defense. The first line of defense consists of external alarms that aim to detect an intruder and sound the alarm before he manages to enter the premises. This line of defense is often susceptible to false alarms because of the movements of birds, animals, innocent passersby, and so on. An alarm raised by this system can only serve to alert you to the fact that something may be happening. It can never provide definite warning that a burglary is in progress.

Perimeter alarms form the second line of defense. They are designed to protect the shell of the building (i.e., the walls, doors, and windows). The perimeter sensors detect the intruder as soon as he breaks into the building. This is the most vital line of defense, because it shows that an intruder has actually entered, or at least attempted to enter, the protected building. Any security guards should now rush to the scene—and remember that if they are far away, they may still be too late to catch the intruder before he gets away. However, the majority of professional security companies still prefer to wait. From their point of view, this is efficient since many burglars deliberately activate the perimeter alarms once or more before they actually enter the building, only to test the reaction of any security guards. However, from the point of view of the homeowner, *you*, this is disastrous. When the burglar finally decides to enter, your security company employees will treat this as merely another false alarm. They will not reach your house in time to prevent the burglary, let alone catch the burglar.

The security companies justify this behavior by the fact that they also install trap alarms, the third line of defense. Trap alarms are detection devices installed on strategic locations inside the house to detect an intruder after he has already entered the building. They are also commonly used to protect individual items of great value or importance. The theory goes that when the security company has monitored first an external alarm call, then a perimeter alarm call, and finally a trap alarm call, it knows that a burglary is in progress. Only then will it react, to avoid rushing men to distant locations because of false alarms. Again, this makes sense from the company's point of view, but you are the one who paid for an expensive security system and guards who can only tell you afterward that, yes, your home was burglarized by some person or persons unknown.

HOW THE ALARM SYSTEM FUNCTIONS

When designing a security system, you must always evaluate the system concept from the point of view of the burglar. Any system may work fine under laboratory conditions, but will it survive first contact with reality in the shape of an experienced burglar? The system may have to cope with deliberate sabotage prior to the burglary, and the system designer will certainly have to consider intruders who attempt to bypass it or in any other way render the security system inoperative. Furthermore, in some places the system must be able to withstand extreme climatic conditions. Cold as well as humidity can affect the functioning of any electronic system. Clearly, the design of a reliable security system is a serious business. Let us go through the functions of a proper alarm system one step at a time. The basic intrusion alarm system consists of the following:

- A control unit, typically (but not optimally) installed within easy reach of the main means of exit and entry, usually the front door. This is the brain of the alarm system.
- One or more warning devices, such as sounder boxes containing warning bells or sirens, or strobe lights, often fixed to the wall on the outside of the house.
- One or more detection devices or sensors.

Other major components include devices for arming and disarming the alarm system, automated dialing equipment, a power supply (such as batteries), and wiring between the various components.

The control unit is housed in a protective box, generally of metal, and is situated in a central location in the area to be protected. An indoor closet might be used for this purpose. It is imperative that the control unit should not be accessible or even obvious to a burglar.

Warning devices of these types are sometimes referred to as alarms or annunciators, even though the word *alarm* is most often used for audible warning devices. Additional devices may be fixed inside the house, mainly as a psychological disturbance to the intruder, and so-called silent alarms are also a possibility. Then a remote signaling system will be used.

The sensor is the device that relays information to the control unit. If the control unit is the equivalent of the human brain, then the sensors are similar to the senses of the human body. A sensor, or detector, is a scanning and screening device. Its effective range is called the detection zone.

The sensors are of different kinds and are, as mentioned above, commonly divided into three lines of defense. The sensors are connected to the control unit, which, after receiving a warning from a sensor, transmits the alarm to the warning device, thus sounding the alarm. The various types of sensors will be detailed in next chapter.

There are three lines of defense but, in fact, four different kinds of alarm protection:

- External alarms
- Perimeter alarms
- Trap alarms
- Deliberately operated alarms

External alarms aim to detect an intruder as early as possible, before he actually reaches the main building. It relies on sensors located in the ground or on the boundary wall or fence. Because these sensors are often susceptible to false alarms, they will generally be monitored by a private security staff and are to be expected only in extremely rich neighborhoods or corporate or government installations. Photoelectric cells are frequently used for this purpose and can be positioned to protect the entire perimeter. Other common sensors used for this purpose are the microwave fence and the field-effect detectors. Special barrier or fence detectors are also used in certain high-risk installations. Another device used for external alarms is the geophone.

Perimeter alarms are designed to protect the shell of the building. The perimeter sensors will detect the intruder as soon as he breaks into the building. Perimeter alarms are very common and are often used in conjunction with trap alarms. Perimeter alarm sensors come in numerous types. Magnetic reed contacts, photoelectric cells, glass-breakage detectors, video detectors, vibration detectors, inertia sensors, infrasound sensors, field-effect sensors, plunger switches, and pressure mats are commonly used sensors in the perimeter alarm. Window foil or prefabricated window foil and wire contacts are also used in some older systems.

Trap alarms are detection devices installed in strategic locations inside the house to detect an intruder after he has already entered the building. They are also commonly used to protect individual items of importance (e.g., a safe). Detectors used as trap alarms commonly include passive infrared detectors, microwave movement detectors, ultrasonic movement detectors, photoelectric cells, magnetic reed switches, pressure mats, light detectors, and video detectors. Other sensors (e.g., field-effect sensors) and sound detectors and heat detectors (not to be confused with the infrared detectors) also fall within this category of alarms. Ionization detectors will also belong to this group, if and when they come into general use. Trap alarms also include special alarms used for guarding valuable paintings, computers, and so on. These alarms can be of many types, but it is common to run normally closed circuits (this is further explained below) incorporated into the main cable to the electronic device or somehow fixed to the object. When the wiring is cut or pulled from the wall socket, whether or not the power is on, the alarm sounds.

Finally, deliberately operated alarms are also known as panic buttons. They are often found in banks but also in many private homes. Such buttons can frequently be found in the bedroom and just inside the front door. Deliberately operated alarm systems are described in a later chapter.

Basically, there are two main types of alarm system installations: self-contained and separate component. Most systems consist of separate parts, but an increasing number of systems are self-contained. A self-contained alarm system contains the control unit, the sensor, and the warning device in the same easily installed unit. Connections on the back of the unit allow the attachment of separate sensors and external warning devices. Both the self-contained and separate-component systems have advantages and disadvantages, so it is wise to plan well ahead when installing an alarm system.

The separate-components systems are highly adaptable and expandable and therefore can be used in any size or type of building; additional components can also be installed at a later time. In their favor, self-contained systems are more easily installed. Generally, they are of the tabletop variety and need only to be positioned in a room and plugged in to be ready to use. Self-contained systems are also easy to move, either from different rooms in the same building or from one building to another. For these reasons, self-contained systems are more commonly used by people who rent their homes or offices, while homeowners and major companies generally use separate-component systems.

There is another, more significant difference between these two types of systems. The sensors used with most separate-component systems are usually designed for perimeter defense (i.e., their purpose is to detect intruders and sound the alarm before the intruders manage to enter the premises).

Older systems thus require an externally mounted key switch that serves as a remote arming and disarming device.

The self-contained alarm systems, on the other hand, are generally designed to detect an intruder after he has already entered the building. Instead of an externally mounted switch, the system incorporates a time delay built into the device that allows a certain time, often 15 to 30 seconds, to enter the building and turn off the alarm. The alarm sounds only if the system is not turned off in time.

It must be remembered that self-contained alarm systems also can be connected to perimeter alarm components. If this option is used, the self-contained alarm functions as a hybrid control unit, both relying on its own sensor and one or more external ones.

Many self-contained alarm systems can in fact also be used in conjunction with separate-components alarm systems, without linking them into the same circuits. The self-contained systems are then used in those areas where it is difficult or impossible to place sensors linked by wire to the control unit in the main system.

Otherwise, self-contained systems are most commonly used in apartments, especially in those where the owner prohibits the tenants from installing permanent alarm systems, but are also used in many different types of buildings. Self-contained alarm systems include the following:

- Passive infrared motion detector units
- Ultrasonic motion detector systems
- Microwave motion detector systems
- Self-contained window or door alarms
- Infrasound detectors

All other types of sensors can generally only be found in separate-components alarm systems.

Most large alarm systems, consisting of more than one sensor, are designed to be broken into zones, whereby different areas of the building are controlled by different circuits.

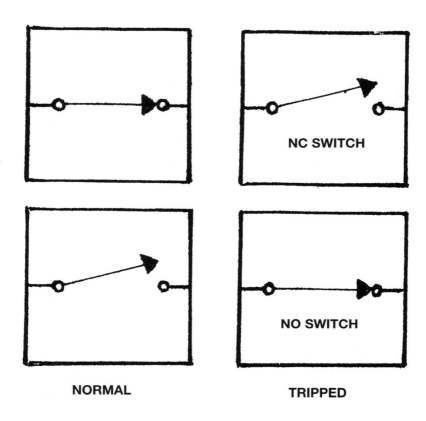

FIGURE 17
The difference between normally open (NO) and normally closed (NC) switches

The main advantage of this is that the occupants of the building can choose to activate all, some, or none of the different parts of the system at any given time. For instance, a bedroom can be excluded so that the occupant can move around there without setting off the alarm already activated in the hall, near the front door.

Each circuit is made up of a series of contacts, all located in the same zone. The alarm will be triggered when any of these contacts is broken. It is common that several different types of sensors are installed on each of these circuits.

When they are designed to be connected to a control unit, all types of sensors operate as simple switches. The sensor switch will be either normally open (NO) or normally closed (NC) (Figure 17). An NC switch is a switch, magnetic or mechanical, in which the contacts are closed (electrically conductive) when no external forces act upon the switch. An NO switch is one in which the contacts are open or separated (electrically nonconductive) when no external forces act upon the switch. The control unit senses when the status of one of its sensors is changing, and then sounds the alarm.

A closed-circuit system is therefore one in which the switches and sensors are connected in series. The alarm is sounded when an activated switch or sensor breaks the circuit, or the connecting wire is cut. An open-circuit system is one in which the switches are connected in parallel. The alarm is sounded when an activated switch closes the circuit, permitting current to flow through it. Open-circuit systems are not as secure as closed-circuit systems.

The alarm system must also contain some means for arming and disarming the system. Arming is the means by which an alarm system is switched on. It may be done either manually, passively, or by remote control. Disarming (switching off the system) is generally done in the same way.

Some self-contained systems are armed and disarmed by a key switch, while a few are armed and disarmed through a code entered from a keyboard.

A passively armed system will arm and disarm itself— for instance, when the correct key is used for locking and unlocking the front door. This will be further described in a later chapter.

Many alarm systems feature a delayed exit/entrance circuit that permits the user to leave and enter the premises without setting off the alarm. The time lapse is fixed, generally between 20 seconds and 2 minutes, and is usually adjustable by the user.

Another common option, especially in simpler alarm systems, is to arm and disarm the system with the help of a key switch (Figure 18). This seems to be especially true of alarm systems in ordinary U.S. houses. The switch is mounted outside the house, and the alarm system does not then have a built-in delay. The lock used in such a system is difficult but not impossible to pick. It is generally a tubular cylinder lock.

Furthermore, the interior of the lock panel has a tamper switch to prevent somebody from defeating the alarm by breaking the device. If the switch is damaged, the alarm will sound. The tamper switch terminals can also be used as a panic button, which is any switch connected to an instantaneously triggered loop. Panic buttons can generally be used whether the alarm is armed or not.

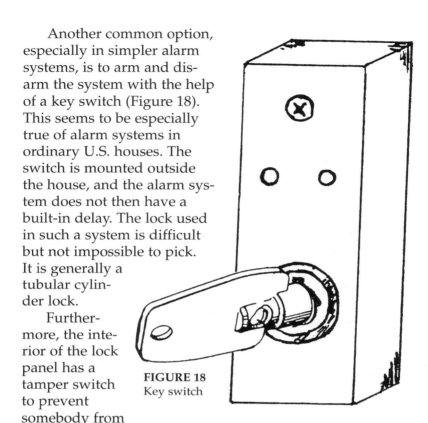

FIGURE 18
Key switch

The tamper switch inside the control panel cover is not the alarm system's only protection against sabotage. A self-actuating warning device has an internal battery inside the sounder box. This will take over as power for the siren if an intruder cuts the wire. Tamper protection similar to the one above is also often employed inside the sounder box.

Other common features to prevent a skilled intruder from disarming the alarm include contacts that are not easily visible on the surface of door or window frames, hidden components of the alarm system, and the use of four-core cable in the wiring. Four-core cable, which is strongly recommended, allows both closed-circuit devices (e.g., magnetic contacts and open-circuit devices, such as pressure mats) to be connected within the same cable. With the former, the alarm sounds if the cable is cut but not if the cable is bridged. With open-circuit devices, cutting the connecting wires will not trigger the alarm, but bridging the circuit will. One pair of wires is used to form the closed circuit, while the other pair is used for the open circuit. The alarm is set off by one or the other of the alarms if the cable is interfered with.

Even if one of these two types of sensors is not connected, many systems allow the other circuit to run into the sensor and then back to the control unit. This protects the system in exactly the same way: it is impossible to sabotage the sensor.

For the same reason, four-core cable is often used together with more advanced warning devices. If the warning device has an integral battery backup, it will operate even if tampered with or if the wires connecting it to the control unit are cut.

Another option is known as line supervision. This is electronic protection of an alarm line accomplished by sending a continuous or coded signal through a circuit. A change in the circuit characteristics, such as a change in impedance due to the circuit's having been tampered with, will be detected and initiate an alarm if the change exceeds a certain level. A normally closed loop is in effect a supervised line.

Control Units

The control unit is the brain of the alarm system and fulfills several purposes. It is used to turn the alarm on and off, and it checks the circuits for faults. The control unit monitors the condition of all sensors and sounds the alarm by transmitting a signal to the warning devices whenever it detects a problem. The warning device is also activated by the control unit if a sensor signals the unit. Of course, the control unit only performs these duties when it is turned on.

The control unit, and by extension the entire alarm system, functions in the following way. The control unit is connected to the alarm switches, which are connected in a loop circuit. The control unit includes circuits for checking the change of status in these switches, thus detecting when a switch opens or closes (depending on the type). This status change will activate the alarm circuit. When the control unit has been activated by a switch, the alarm system is said to be tripped. This activates and sounds the alarm.

Loops of alarm switches are connected to the input terminals of the control unit. There are generally separate terminals for connecting a number of switches, or loops, to the control unit. Each terminal identifies a different loop of switches in the system. There is a separate loop for every single easily defined area in the building protected by the system. For instance, one loop might protect the front door, while another loop protects all windows, and so on. In this case, the front-door loop probably has a built-in delay, since this is the way to enter the house, and the owner must have some time to disarm the alarm system. But the other loops of the perimeter defense are triggered instantaneously because these areas would be used only for entering and exiting the house by an intruder.

There might also be what is generally known as a day loop. This is also triggered instantaneously, but will usually activate only a small buzzer instead of the main warning devices. The day loop is sometimes used by families with young children and warns if the children leave through the front or back door or open the gate leading to the swimming pool. This is not really a part of the intruder alarm system, although to a certain extent it uses the same circuits.

The control unit and the entire alarm system are said to be armed, or set, whenever the control unit is in operation. Otherwise the alarm system is disarmed.

Although control units with more complex circuitry also

exist, this is the basic configuration of a control unit. Other features do not affect the basic functioning of the alarm system much. Such features concern only certain aspects of the system. For instance, some systems might turn off the warning device after a certain period of sounding the alarm, often five minutes, and then shut off the alarm and reset the system. Other control units assign priorities when signals are received from more than one loop in the sensor system. For instance, the loop used for fire or smoke detectors will generally override all other loops. Incidentally, this can be turned to the advantage of the intruder if the protected area is large enough. If a fire is started in one part of the complex, other parts of the building might effectively be prevented from sounding the alarm during a break-in. For this reason, it is advisable not to combine burglary and fire alarms.

As can be seen from this description, a complex system includes several sensor loops. The really critical connection, consequently, is not between the control unit and the sensor loops but between the control unit and the warning devices. If that line is cut, no warning will be sounded. A skilled intruder will attempt to break this connection rather than futilely trying to avoid the sensors, which is often almost impossible to do if the alarm system is designed properly.

The typical control unit has both instant and delayed loops. An instant loop will sound the alarm immediately when a switch is tripped, while a delayed loop will wait for a certain period. The entire system usually has a delay function, at least in modern systems. This allows sufficient time to arm the system on the way out and disarm the system on the way in, as well as obviating the need for an external switch.

Most control units allow different loops to be turned on and off separately. This is useful for guarding the entrance at night, even if the owner is sleeping in his bedroom, and allowing the smoke or fire alarm to function at all times.

The number of sensors connected to one loop is not limited, even though there is only one set of terminals per loop. The switches are merely wired in series or in parallel (Figure

19). NC switches are wired in series. When the switch opens, signifying that a window or a door is open, this breaks the NC actuating circuit in the control unit and the alarm sounds.

NO switches, on the other hand, are wired in parallel. When any of the switches closes, signifying that an intruder has been detected, this completes the NO actuating circuit in the control unit, and the alarm sounds. An NC switch can be bypassed by bridging the circuit. An NO switch can be bypassed by cutting the circuit.

A loop of NC switches is generally more efficient because it is being monitored continuously by the control unit. Any interruption of the NC loop will cause the control unit to sound the alarm. This is useful if, for instance, the wiring connecting the switches is broken by an intruder. The NO loop can be easily cut, and the control unit will then not notice the difference. Remember, though, that a loop of NC switches is often run together with the NO loop only to serve as a sabotage alarm.

In addition to this sabotage alarm, most professional control units include a plunger switch, known as a tamper switch, which detects when the door of the control unit has been opened. Because the tamper switch is on an instant loop, the alarm sounds as soon as an intruder tries to disable the system by damaging or turning off the control unit.

Also, the control unit is often fitted with an additional tamper switch, installed in the back of the control unit. This switch sounds the alarm if the control unit is removed from the wall. These two types of plunger switches are NC.

Since the control unit is often located in a cabinet, this door might also be fitted with an additional sensor that helps to guard the control unit, sounding the alarm when the cabinet door is opened.

The control unit is most often placed in an accessible but not easily seen place (e.g., a closet in the living room, hall, or bedroom; the kitchen pantry), since it must be reached frequently to arm or disarm. It is, however, better to use a remote switch. Such a switch is easier to hide, and the intruder cannot neutralize the control unit, thereby deactivating the security system, merely by finding the switch. Either an exterior switch or an interior switch may be used, but an interior switch is most often located in one of the accessible places or near the main entrance.

The control unit must always be located in a place that can be temperature-controlled—usually not in the attic, garage, or basement. The unit should be firmly attached to the wall, so that it cannot be removed easily without both damaging the wall and the control unit and, of course, triggering the alarm. Despite this, destroying the control unit will generally render the alarm system inoperative, even though the alarm may sound once.

The actual control unit is sometimes completely hidden, while the status of the system is shown on a separate control panel. This panel, even if tampered with, cannot control the system because it only gives information of the current status.

Modern alarm systems are often connected to the mortise lock in the front door; the system is disarmed whenever the door lock is opened by key. Picking the lock will produce the same result.

You should try to understand how your control unit works. Not all types and settings are tamperproof. One interesting way to disable certain control units is to remove the jumper wires between the unused NC connection terminals. In effect, this opens the switch, and, because the control unit senses an open circuit, the alarm will go off. Such minor tampering can give the impression that the alarm system is faulty, maybe making you turn it off until a maintenance man arrives, who will of course understand the real problem. Meanwhile, your expensive security system is inoperative. Guard against the possibility of an electronically skilled intruder's getting to your control unit if he manages to visit inside your home on one pretext or another.

Power Supply

Most alarm systems are powered through the control unit.

Generally, the power supply is housed in a separate transformer box plugged directly into a wall outlet. The transformer converts the alternating current (AC) to 12 volts direct current (DC), which is used for the entire alarm system. The transformer is attached to the control unit by an ordinary two-conductor wire.

If the intruder can gain entrance to the building legitimately (e.g., during office hours), he can sometimes locate the transformer and simply remove it from the wall outlet. Since most alarm systems have a battery backup power source, the alarm system will switch at once to battery power (and usually without alerting the inhabitants in the house). This is generally enough to last for several hours, but the power will run out soon after the occupants have left the building in the evening or gone to bed at night. This system-disarming method is often helped by the fact that most wall outlets are located low on the wall and in inconspicuous locations, such as behind furniture. Even though the control box might be checked in the evening, the transformer will most often not be checked.

Backup batteries are commonly used because burglars have learned to shut off the electricity in the houses they enter. The backup batteries are rechargeable and usually installed in the control unit. The rate of discharge depends mainly on whether an audible alarm has been activated or not—this consumes a lot of power—but the backup protection will generally last for three hours or less if the alarm is sounded.

It should be noted that especially in Europe the backup batteries are more often designed to last for at least 24 hours or even for 72 hours (a normal weekend). Do not buy cheaper batteries.

If the alarm system has no backup batteries, the system will always come back on armed if the AC power has been cut off temporarily.

Some control units check the status of the batteries each time the system is armed or disarmed. This is to inform the user that power is low. Although it is more expensive, this is a highly recommended feature. How many of us would remember otherwise to check the batteries from time to time? Without reliable batteries, an expensive security system is only a pile of junk.

You should also test what happens to your system when power suddenly returns after a power failure: Will the system remain armed, or will it malfunction? Needless to say, good security systems should be able to cope.

As a point of interest, it should be mentioned that sometimes there are external control panels installed to inform owners of the current status of the alarm system. These also alert owners that the current required to recharge the backup batteries is insufficient, which is to let owners know that they must check and possibly repair this function. Opportunist burglars also like this information: it tells them that it is safe to break into the house as long as they cut the power first. Avoid such control panels on your system.

Warning Devices

Most alarms encountered are local alarms (i.e., alarms that when activated either make a loud noise at or near the protected area, flood the site with light, or both). The alarms may also be so-called remote alarms, which transmit the alarm signal to a remote paging unit or monitoring station. The alarm might then be a silent alarm and not give any obvious local indication than an alarm has been transmitted. Sometimes a combination of local and remote alarms may be encountered.

Warning devices come in three basic shapes and in numerous types. All of them are suitable for installation in an alarm system, and any number or combination of warning devices can be used together. Most common are low-voltage sirens or bells, but regular klaxons are also in fairly frequent use.

The sirens are designed to emit a distinctive warbling high/low tone that can be heard over a long distance. In some alarm systems, a siren is used to indicate a burglary, while another siren or a mechanical bell, emitting a steady tone, is used to indicate a fire. An ordinary bell is seldom used any more as an intruder warning device. A buzzer may also be used, but this too is a dated warning device now used primarily indoors or in factories.

The electric siren is the most common warning device today. A number of different types are in frequent use, but the piezoelectric siren is characterized by the lowest power requirements and is therefore often recommended. The warning device must usually be connected to a battery in case of power failure, so a low level of power use is advantageous.

The siren in an intruder alarm system is invariably of the rise/fall type, used in most European countries as a police siren.

The design of the siren will not always be the same, however. Some sirens are designed for outdoor use, others for indoor use only. The difference is not only in the level of protection against the weather that its cover affords the interior circuits; the emitted noise of these sirens is also quite different. The aim of the outdoor siren is to be heard for as far as possible, whereas that of the indoor siren is to scare away the intruder by emitting an ear-splitting noise. Sirens emitting a low-frequency sound have the longest range and are used outdoors. Sirens with high-frequency sound are used indoors because they are a very effective psychological deterrent to intruders. So professional intruders performing quick break-ins without bothering to disarm the alarm systems occasionally use ear protection.

Yet another warning device, similar to a siren, is the pneumatic warning device. Because this relies on compressed air, it does not need much power. The compressed air, controlled by an electric air regulator, produces a very powerful sound when released.

Most types of sirens will produce a sound of at least 100 decibels or more, and sound levels of up to 136 decibels are not uncommon.

Another type of warning device is the so-called silent alarm. This is an automatic dialer that places an emergency telephone call to somebody, either the owner, the local authorities, or someone else, and then plays a taped message indicating the problem, whether it is a fire or a break-in. Telephone automatic dialers and other remote signaling systems will be detailed in next chapter.

Some alarm systems, especially those without a remote key switch, also use a small pre-alarm buzzer or piezo sounder that will sound during the entry delay time to remind the owner to disarm his alarm system before the real alarm is sounded.

Finally, there are strobe lights. This type of warning device is especially useful in a crowded neighborhood, where it might be difficult to pinpoint the origin of a siren. The strobe light, if one is installed, is often the weak link in an alarm system. The light is connected in parallel with the siren or bell used as a warning device, because it is generally only an auxiliary system designed to scare away the intruder and help pinpoint the source of the warning sound from the siren.

This means that the strobe light is electrically connected to the same contacts in the control unit as the siren, and so the entire alarm will be rendered useless if an intruder first inconspicuously short-circuits the strobe light. This can even be done with water if the strobe light is not sufficiently weatherproof or provided with a tamper switch. And, of course, the wire can be cut.

For this reason, the wiring is usually hidden or at least out of reach. Even so, it's remarkable that so many alarm systems are easy to disable by simply cutting the wires to the siren or to another warning device. You should ensure that your system is not this easy to disable. After all, what does it matter that your expensive control unit functions perfectly if it has no way to sound the alarm?

The warning device, whether a siren or strobe light, will often be mounted near the eave or roof line of the building, at a position at least 2.7 meters high. It might include a tamper switch, and the wiring will generally be through the attic. Needless to say, it is a favorite tactic of professional burglars to first break into the unprotected attic to locate and cut the wiring to the siren, and only then break into the house.

Sometimes the warning device is mounted on a wooden or metal pole, such as a television antenna. It is not uncommon for it to be mounted directly on the already existing mast of the television antenna. This is especially true of strobe lights.

Many, but not all, warning devices come with a tamper switch, allowing the alarm to sound even if the siren is tampered with. Of course, this is worthless if no other warning device is present: a clever intruder will look out for an extra hidden warning device before he disables the main one.

It should be noted that many sirens can be silenced by spraying them with foam (e.g., insulation foam), which is often an easy way to silence an otherwise functioning alarm system. No tamper switch will help in this case, so always aim to use more than one warning device and position both or all where they cannot be seen or reached easily.

In most countries, the sounder box turns off automatically within a preset time, usually no longer than 20 minutes, so as not to disturb the neighbors too much in case of a false alarm. The system then automatically resets itself. Now, this might not always be the case with government installations. Even if the noise is turned off, there is often an affixed strobe light that keeps flashing until someone manually resets the alarm.

Remote Signaling Systems and Automatic Dialers

Many alarm systems are linked to a telephone. By using a separate, ex-directory telephone line only used for outgoing calls they are designed to raise the alarm in a remote location in case of intrusion. These devices, known alternatively as automated dialing equipment (ADE), automatic dialers (or autodialers), or telephone dialers, are available in three slightly different types.

The reason for using an ex-directory line is that the alarm cannot be neutralized by calling the automatic dialer. Placing such a call effectively will block the outgoing call. In some countries, the telephone system allows all incoming calls to be routed to a different number, thus freeing the automatic dialer's line.

The simplest type of automatic dialer is programmed to dial the local police and then play a standard prerecorded message by phone to the police, telling them that there is an intruder in the house. The message includes such relevant details as the owner's name, address, and phone number. Although in theory reliable, this system is frequently useless, because many police forces no longer have enough personnel to monitor the lines. Another drawback is that the automatic dialer will not work if an intruder cuts or temporarily disconnects the telephone lines. Furthermore, the automatic dialer generally dials only once. If nobody answers, it's too bad for the owner of the system.

Another version of this system works in the same way, but instead of calling the police, the automatic dialer alerts a neighbor, friend, or relative, who then calls the police. This system is even worse—the friend might not be at home, and, even if he is, the call will take longer to reach the police. Here, too, a cut line prevents the system from sounding the alarm.

For these reasons, the digital communicator is more popular. This is a more sophisticated system that is able to dial a central monitoring station, usually the security company's central control station or the main security station in a corporate complex. Here personnel are constantly on duty to watch annunciators reporting on the condition of the alarm system. The alarm call consists of a series of coded signals that comes up as text on a computer screen, monitored 24 hours a day by staff members who immediately alert the police—and often dispatch their own security team to the location.

The digital communicator continues to call until the message gets through. This is determined by the receiving station's giving a correct code that means the message has been understood. In this case, the digital communicator might be programmed to call several different numbers until a confirmation has been received from one of them. Other systems are designed to confirm by calling back within a specified time. Generally, this system is designed to also register any faults on the telephone line, thus alerting someone to the danger of a cut line. This system also relies on an ex-directory line.

Locations guarded by such systems are generally easy to recognize: the company that did the installation will advertise the system's presence by posted signs. Systems of this kind are almost always rented as parts of professionally installed alarm systems and are subject to regular maintenance inspections by the security company.

An even more advanced system, marketed in Britain and some other countries, also responds to a fire alarm. In addition, the system indicates in exactly which zone or loop the alarm has been triggered and what type it is. Any faults either on the line or in the alarm system are also reported.

The most exclusive alarm systems always use direct private lines rather than the ordinary telephone lines. The alarm signal is transmitted on a continuously monitored private line to the security company's central monitoring station, and any fault or interference with the line is noticed right away. Such a system is used only in high-risk installations or in the homes of the extremely wealthy.

If an ex-directory or a private line is not being employed, the automatic dialer must always be connected as the primary telephone in the house. This means that all incoming calls must pass through the automatic dialer and then go to the regular telephone. If this is not the case, the system, even if it is hidden, can be easily disabled by removing the regular telephone's handset from the hook.

Some automatic dialers have their own built-in backup batteries, allowing the device to work for several hours even when the power is cut. The power ordinarily comes either from the control unit or a separate AC power adapter.

The recorded message uses either a magnetic tape, in which case the owner can record his own message, or a prerecorded computerized voice. Most automatic dialers can be programmed with up to three telephone numbers, all of which will be called in order.

An interesting point about certain of these systems, especially the British ones, is that some police forces insist that the siren, if one is used in conjunction with the remote signaling device, have a built-in delay, so that it will sound the alarm three to five minutes after the message has been relayed to the police. This is to give the police a greater chance of catching the intruder in the act.

It should also be noted that even if a security company is connected to the automatic dialer, it is by no means certain that its personnel will care to respond to a single triggered alarm. False alarms are now so common that most companies of this kind will wait until they have first received a signal from an external detector, then a signal from a perimeter detector, and finally a signal from an interior detector, indicating that this is indeed a serious intrusion attempt. Only then will they dispatch patrols or call in the police. This aversion to responding to false alarms may give the intruder a few valuable minutes in which to do his work and get away.

A radio transmitter can also be used as an automated dialing system. In this case, it is most common to use a transmitter in the frequency range of around 27 MHz, because this is commonly used in personal paging systems. Using a radio system this way is most common in advanced alarm systems in vehicles, boats, and other structures where there is no regular telephone connection. The suitability of the antenna is the most important factor, because this will determine the range of the system. If the antenna is removed or covered by a metal box, the transmission will suffer a severely curtailed range or even disappear completely. Naturally, no remote alarm will sound.

Access and Exit Control Systems

Access control is the means by which only authorized persons are allowed to enter a building or flat, while unauthorized persons are kept out. Such a system is commonly designed around an exit-entry control system or a system to arm and disarm the alarm system. There is an authorized access switch that makes all or parts of an alarm system inoperative in order to permit authorized access.

In most doors the mechanical lock is the only access control system. However, there are also many other possibilities, mechanical and electrical, sometimes used only as keyless locking devices, but sometimes used together with alarm systems. In the latter case, the system definitely relies on electrical control switches.

Most of these control systems require the use of combination codes. (See Chapter 9 for more information on this.)

Electrically operated locks are becoming more and more popular, especially in industrial installations and offices. Electrically actuated release catches are so far more common than pure electronic locks, however. These units operate on low voltage, often 24 volts: for this reason they need transformers. Locks that are normally locked will be unlocked when the system is energized. On the other hand, normally unlocked locks will lock when the unit is energized. The first option is most common.

It is sometimes possible to enter by connecting a high-power source to the lock, thus overloading the circuit. However, some locks of this type are so-called fail-safe. This means that they will automatically unlock if the power fails, such as might happen in a fire. This is meant to provide for a safe escape route, but it can also be used for purposes of gaining entry to the premises. In short, tampering with the power supply might well open a lock of this type, as long as the intruder knows what type of locking device he is dealing with.

Also note that there might be a considerable distance between the actual lock and the remote control unit. Most common, of course, is to have the control unit next to the door. The control unit is always built around a control switch. Remember that it is preferable to use a remote control switch, away from the hidden control unit.

There are numerous types of control switches, including digital access control systems, electronic or mechanical card access control systems, lock switches, remote control switches, delayed alarms, and ordinary key switches. The various systems may rely on number code combinations, coded cards, or plastic keys instead of metal keys. All are common in hotels and many hospitals. Because these systems quite often are electronic, they are frequently connected to a registration unit that is able to record when a certain code or key is used and where if the system includes more than one lock.

The digital access control system is perhaps the most common. It usually consists of two parts, an access-control keyboard and a program unit. Such a system can use up to 10,000 different code combinations. Every legitimate user might have an individual code, or the same code might be used by everyone. The code might be used for opening the lock as well as for disarming and arming the alarm system.

Less commonly, a dial of the type found in combination locks is used in this device. Both types are very popular in major companies, which might find it necessary to change the code from time to time (e.g., when staff members are replaced or keys are lost).

Some digital systems close the circuit only momentarily, thus turning off the alarm system/or opening the door for up to 10 seconds or so. Others use circuits that remain closed until the code is reentered, thus remaining unlocked for the period in between.

Most of these switches only disarm one loop, consisting of the sensors that guard the nearest way to the control unit, so that the owner can immediately proceed there to disarm the entire system.

The code usually consists of either four or six digits. Occasionally the four-digit code serves as a code lock, while the full six digits arm or disarm the alarm system.

Sometimes the lock does not unlock until the alarm system is disarmed. It is also common for the code lock to block another attempt temporarily if the entered code is the wrong one—this discourages attempts to enter by using random combinations. Some systems go even further and trigger an alarm when a preset number of incorrect combinations have been entered.

Most systems of this type also have a preset timer that is activated when the first digit of the code combination is entered. If the remainder of the code is not entered before the time expires, the entry is canceled.

Yet another interesting feature of some of these systems is what is known as a duress alarm, which is connected to an alarm system and a warning device (e.g., a siren). A silent alarm is also a real possibility here. The duress alarm is activated by punching in the correct combination code but replacing the last digit with a predetermined other digit, the duress digit. This will or will not open the lock, depending on the programming, but will definitely trigger the alarm. This is to warn against entry made under duress, for instance by an employee who is held at gunpoint by an intruder. Technically, this system is a momentary switch with NO switch contacts.

Another unit, very often used in conjunction with this device, is the card reader. A card-based access control system is easy to use and therefore popular in large corporations and government installations. The card reader is situated next to the door, requiring the person wishing to enter to both insert his coded access card in the reader and also punch in his personal code on the keyboard.

The plastic cards for these locks are of two different types. They can either rely on a magnetic code, in the same way as a credit card, or they can have various numbers or figures for optical reading. In either case, the card must be inserted into a slot or passed through a card reader situated at the door. There are also systems in which the card reader is invisible, hidden in the wall next to the door, and the card is simply displayed at around 30 centimeters from the reader in the wall.

Whatever type is being used, the card has a code that allows entry through only one or several doors, depending on the design and programming. All cards can have the same code, or individual codes might be used instead. The system is extremely flexible.

A coded plastic key can be used. It works exactly like a card key but is designed to be carried like an ordinary key on a key ring.

In certain of these locks, the system is not electronic but mechanical, even though the key is still replaced by a plastic card. A plastic card with holes in it means that the lock is definitely a mechanical one, the holes in the card fitting exactly to a number of balls within the lock. The lock can be individually coded, and the code can be easily changed. The card may have a code allowing access through several doors or only through one. Note that this is not a real code, merely a means of ensuring that the card fits into its slot to open the door. The holes in the card function the same as the cuts on a regular key.

A door protected by a code lock or a card lock often has a sensor to sound an alarm if the door remains open too long. Sometimes this is only a buzzer to indicate that the door must be closed, but occasionally a real warning device will be used.

Electronic locks might also take advantage of time coding. This system is programmed to allow or deny access depending on the time of day or night. Every legitimate user might be allowed to enter during ordinary office hours, but only key personnel might be permitted to enter after office hours. When a system of this complexity is used, an automatic registration unit almost certainly exists. A system of this type can be made very flexible.

If a card is lost, the code (if one is being used) should be changed immediately, and the person who lost the card is simply issued a new card and a new code. Changing the code is a quick process; the number of possible combinations is very high.

It is difficult to bypass access control systems of this type. The keyboards and card readers might be vulnerable to weather, but this does not allow access; it only further prevents it. Off course, a few ways to cheat the system exist. You should be aware of these possibilities and ensure that they do not occur.

If the same code has been used long enough, wear and tear and dirt will sometimes indicate which keys are most commonly used. This doesn't reveal the combination itself, but does decrease significantly the number of possible combinations. This has often helped an enterprising intruder to gain entrance.

Another common way of learning the code (at least in those locations where no card is required) is simply to observe, from a distance, somebody entering. For this reason, it is a good idea to fit your keyboard with a protective cover.

Electric digital code switches can sometimes be shorted by connecting a high-power cable to them instead of the low-power kind normally used. This might open the lock, but only if it is of the correct type, the one that will unlock when energized.

In private homes, the access control is usually simpler. One common access control system is the lock switch. This device has two purposes: to disarm the alarm system and to unlock the door. The system is usually installed in the mortise lock in the front door (see next chapter). This is common in private homes, where the number of keys is limited. When this system is used, the door and its frame are often protected by an inertial sensor.

Remote-control switches are another possibility. They come in two major types, the most common of which being the radio-transmitter-operated switch, followed by infrared-operated switches. This system is frequently used in garage doors. The remote control opens the door and disarms the alarm system at the same time. Remote-control switches, especially of radio-frequency type, are common in perimeter alarm systems around outlying sheds and stores. The alarm is built around sensors mounted on a surrounding fence, but the remote-control device can arm and disarm the control unit located inside the shed or store. The range is around 70 meters, which is usually enough.

The radio transmitter transmits a digitally coded signal. This signal, when recognized by the receiver switch, arms or disarms the alarm system or only parts of it, if so desired.

Also common are delayed alarm systems. In these the alarm is simply delayed, so that the operator has enough time to reach the control unit and turn off the system before the alarm is sounded. (The delayed alarm was described in more detail above.)

It should be pointed out that some corporate alarm systems are not armed and disarmed manually at all. Instead, they rely on a timer switch that turns the alarm system on and off after and before office hours. Note that the time will not necessarily be the same every day of the week.

Finally, a key switch is exactly the same as a standard lock but connected to the alarm system. It is usually located in a small box, protected against sabotage by a tamper switch, and located outside the building. Because the switch can sometimes be tampered with by exposing the wires leading to the key switch, it is often at least set on a hardened surface (e.g., concrete) and is usually protected by an inertia sensor or other device.

If at all possible, design your security system so that it can be used even if the access control system has temporarily broken down. Many burglars sabotage the access control system before they actually break in, so that the alarm has sounded and the owner forced to shut the security system down while waiting for the repairman.

More expensive than the electric lock is the electromagnetic lock, which is used in hospitals, banks, prisons, airports, and numerous other high-risk installations. The interesting thing about the electromagnetic lock is that it can exert a very strong holding force. The lock consists of two components, the lock itself and its armature (Figure 20). The lock is mounted on the door frame, the armature on the door: both components are designed to be sturdy and resistant to physical abuse. The lock and armature make contact when the door is closed. Locking, or activating, the lock causes the two units to be magnetically attracted to each other and hold together. The wiring is factory made and includes tamper-resistant circuits.

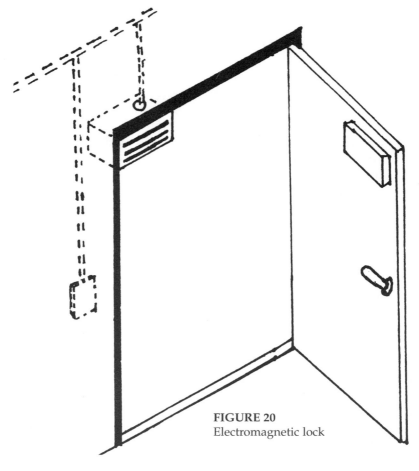

FIGURE 20
Electromagnetic lock

The easiest way to disable the electromagnetic lock is to deprive it of energy. Make certain that this will not be easy for an intruder trying to get into your home or office.

Lock Switches Built into Standard Mechanical Mortise Locks

A lock switch can also be built into a standard mechanical mortise lock. The lock switch again fulfills two purposes: to disarm the alarm system and to unlock the door. The standard key to the lock is used, and while the key will unlock the lock, the switch will disarm the alarm system. The system is usually installed in the mortise lock (Figure 21) in the front door, although other secure locks, such as tubular locks, might also be used for this purpose.

This device is common in private homes, where the number of keys is limited. When this system is used, the door and its frame are often protected by an inertia sensor or some other device. It is possible to pick this lock, but an even greater danger to the owner is having the key to the lock fall into a burglar's hands, since this not only will allow entry to the building but also will disarm the entire alarm system.

The lock switch is a standard contact switch, which is mounted inside the lock next to the latch bolt. The switch will react when the latch bolt is moved to the locked or unlocked position. Such a switch is in itself relatively easy to manipulate, so other means of protecting the switch are also usually installed.

One efficient protection system uses a mortise lock together with an inertia sensor built into one of the jumper cable terminals mounted on the door frame (Figure 22). This sensor protects both the lock and the door. There is also an light-emitting diode (LED) in the face of the lock. The LED not only indicates that the alarm is armed by flashing, but also protects against sabotage because its circuit will trigger the alarm if it or the lock is damaged. Furthermore, if the power is not sufficient, the LED will stop flashing and remain on continuously. This also happens if the alarm has been triggered. The system will be automatically reset every time the door is locked.

The switch circuit is an NO system, open when the lock is locked and closed when the lock is unlocked. The wire is usually of the four-conductor type, however—two conductors are the NO circuit; the other two are the NC defense against sabotage. The wiring is run either through a diagonal hole in the door or on top of the door.

An alarm of this type can also be used if a number of doors are to be used for main entrances. The locks and their systems are connected in parallel. The alarm system will be disarmed whenever any of the locks is unlocked—however, the system will be armed only when the last unlocked lock is finally locked.

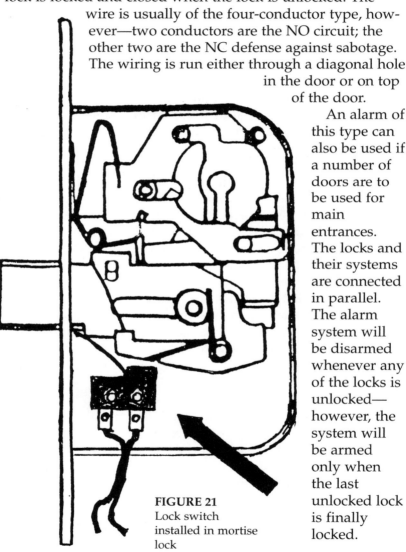

FIGURE 21
Lock switch installed in mortise lock

Wiring

Depending on your type of security system and how you install it, the wiring may either be the easiest way to deactivate your system or its most secure feature. This is undeniably a question of how much you pay for your system.

Most wiring connecting the various sensors to the control unit should be concealed, or at least positioned very unobtrusively. The wiring is frequently run inside the interior walls, with a small hole drilled near the sensor. The wiring goes through this hole into a larger hole in the sole or top plate in the basement or attic. Run through there, the wiring follows the attic or basement until it goes into and comes out of a small hole next to the control unit. Make sure that a burglar cannot locate and tamper with the wiring if he breaks into the basement or attic.

Wiring can also be hidden under carpeting, floorboards, and masonry trim or behind furniture. It is generally not placed where it might get damaged by water, excessive moisture, or local pests (rats and other rodents sometimes bite through electrical wiring).

Keeping in mind that a loop of NC switches is often run together with the NO loop in a four-strand cable to serve as a sabotage alarm, a security system should not be made inoperative by simple short-circuiting. The most inconspicu-

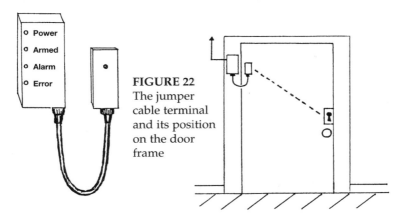

FIGURE 22
The jumper cable terminal and its position on the door frame

ous defense against sabotage is the balanced alarm system, which is impossible to detect from the outside, or even from the inside, of the building as long as the control unit is not found and opened.

A balanced system consists of a number of sensors (as does an ordinary system), but every sensor is fitted with a unique resistor. The control unit is adjusted to recognize the total resistance of the system. If one sensor is shorted or bypassed, the control unit detects a decrease in the system's total resistance and immediately triggers the alarm.

It should also be remembered that some contemporary alarm systems use built-in computers to control the system. This means, among other things, that any changes in the system's status will be recorded, even if the alarm has not been sounded.

More advanced alarm systems employ computerized communication via two or four wires to further protect the system against sabotage. The computer queries the various detectors about their status. The entire net of wiring will also be monitored in the same way. Because this system is almost always connected to a central alarm, the entire system (including any detected irregularities or triggered alarms) can be displayed on a computer monitor. One example of this system is the U.S. Vindicator system, originally designed to protect air bases and important defense industries but now used by major private companies as well.

In this system, every detector is connected to a transponder that reports through time-multiplexing every change in wiring or any indication given by the detector. Reporting is in real time to an alarm operator in a security-monitoring unit. This system is extremely difficult to get around. With this one, it is definitely easier to subvert the operator, because a human being is the weakest link in the chain.

Wireless Systems

Wireless alarm systems are frequently used. Even though they are more expensive, they are easier to install than wired systems. This kind of alarm system consists of a central processing unit and one or more detection devices. All the components communicate by radio wave. The radio frequencies used are supposedly free from interference from any other radio-controlled equipment, as well as from communication units in passing taxis and police cars.

The entire alarm system is controlled by a small radio touch pad about the size of a pocket calculator. This remote control can be used to activate the alarm from anywhere in the house. It can also be used as a personal-attack alarm or panic button wherever the user is, whether in the house or the garden or garage.

In the United States, some wireless alarm systems are available that not only sound the alarm but also regulate turning on and off lights and other electrical appliances. These systems are supposed to scare away intruders by mimicking the presence of somebody in the house, even when it is empty. As we have seen above, experienced burglars are not fooled. This system is either battery powered or takes its power directly from the AC power supply. In the latter case, the entire system will fail if the power is cut. Avoid such high-tech toys if you value your security.

If batteries are used instead, the battery-powered transmitters will be supervised every 90 minutes or so to determine battery and functional condition. This is a safeguard against loss of power, but only as long as the control unit itself remains powered, of course.

A wireless alarm system must not be confused with a self-contained alarm system. The latter also relies on radio sometimes but works in a different way. A real wireless alarm system is, in effect, a separate-component system that uses radio waves instead of wiring.

CHAPTER 3

Alarm Sensors

There are many different types of sensors, which are often divided into active and passive sensors. Active sensors create a field and detect a disturbance in that field, while passive sensors detect natural radiation or radiation disturbance without themselves emitting the radiation on which the sensor's operation depends.

Sensors are also divided into categories depending on the area or particular point that they protect:

- *Perimeter protection* is the protection of access to the outer limits of an area by means of physical barriers, sensors on these barriers, or external sensors not associated with any physical barrier.

- *Interior protection* is a line of protection along the interior boundary of a protected area, usually a building, including all points through which entry can be made.
- *Area protection* is the protection of an inner space or volume of a secured area by means of a volumetric sensor, or a sensor with a detection zone that extends over a volume, such as an entire room.

MAGNETIC REED SWITCHES AND WIRE CONTACT SYSTEMS

A magnetic reed switch is an alarm system detection device in which a disruption of the magnetic field between two points causes a break in the electrical current. This break signals the control unit to activate the alarm. Magnetic reed switches (Figure 23) are among the most common alarm sensors in use today.

The magnetic reed switch consists of electrical contacts formed by two thin, magnetically actuated metal reed-like vanes, held in position (NO or NC) inside a sealed glass tube. The tube is enclosed in a metal or plastic case. The device

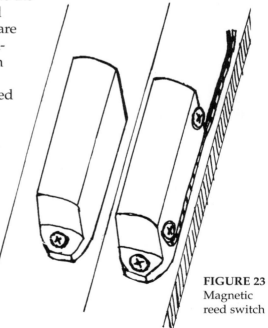

FIGURE 23
Magnetic reed switch

works on the principle of magnetic attraction. The sealed metal contacts are positioned in such a way that, when a sufficiently strong magnetic field is present, they are either pulled together or pushed apart (Figure 24). When the magnetic field is removed, the metal contacts will naturally move in the opposite direction.

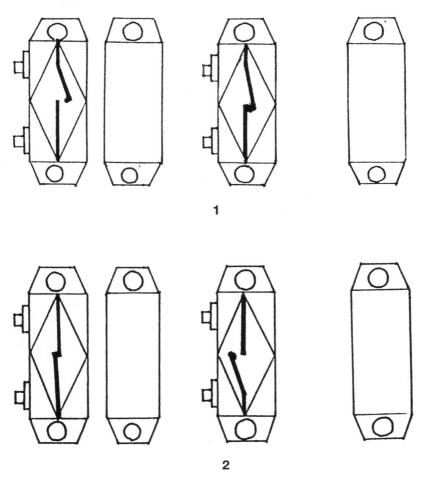

FIGURE 24
NO (1) and NC (2) magnetic reed switches

The reed switch is composed of two separate units: the magnetically actuated switch and a large magnet enclosed in a similar plastic housing.

When the magnetic field is strong enough, the contacts inside an NO reed switch do not touch, so the loop are open. But when the magnet is removed, the contacts are pulled together, thereby closing the contacts. The opposite is true of the NC reed switch, in which the contacts are closed when the magnetic field is affecting them. If the magnet is removed, the contacts will open.

The magnetic reed switches are set into the doors and windows in a zone, and are surface mounted or recessed. Recessed switches (Figure 25) are invisible when the door or window is closed, because they are set into recesses in the frame. So, they are not as easily tampered with as the surface-mounted reed switches. The circuit is broken whenever the door is opened, and this triggers the alarm.

The switch is usually mounted in a fixed position, such as on a window frame or a door jamb, opposite the magnet, while the magnet is fastened to the window or door. When the window or door is opened, the removal of the magnet

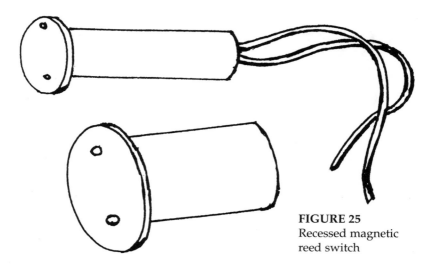

FIGURE 25
Recessed magnetic reed switch

forces the switch contacts to change status. This will open or close them, depending on the type of switch. In either case, the alarm will sound.

NC reed switches are the most common because they are the most secure. NO reed switches can also be used, but, again, they can then be circumvented by simply cutting their wiring.

Magnetic reed switches typically are mounted as far away as possible from the hinges, so that they will trigger the alarm even if the door or window is only partly opened. The switch can be located near the hinges only if the owner wants to open the window at night for ventilation. An ordinary reed switch is generally triggered when the magnet is moved away more than about 2 or 3 centimeters, although some switches, so-called wide-gap switches, allow the window to be opened 5 centimeters or so. These latter switches are used when the construction of the window or door doesn't allow the two switches to be mounted closely together.

Reed switches come in different shapes and brands, but they are typically about 3 to 4 centimeters long and a little more than a centimeter thick and wide. For obvious reasons, they are always mounted in pairs: the one housing the magnet (always the one on the moving part of the door or window); and the other, housing the stationary switch part (mounted on the door jamb or window sill). When the window or door is closed, the two parts will be very close to each other. A burglar will attempt to locate the wiring, even though it is sometimes hidden in the wood. The terminal screws for the wiring will generally be visible, however.

Magnetic reed switches are not very safe. A sophisticated burglar will deal with a reed switch in the following way if he suspects that one is present. First, he'll spend a few seconds locating the reed switch with the help of a small pocket compass, which will also indicate the position and polarity of the reed switch magnet's magnetic field. The burglar will then fasten his own sufficiently powerful magnet in the location that corresponds to the original magnet's position and polarity. He can then open the window without fear of the reed switch's sounding the alarm.

Alternatively, he may attempt to drill a hole in the window frame to short-circuit the wiring of the reed switch.

Some alarm systems also include miniature reed switches. These work in exactly the same way as ordinary reed switches but must be properly aligned and also very close to each other, because the magnet is smaller and, consequently, the magnetic field not as strong. Reed switches can sometimes be found in other locations (e.g., cabinet and cupboard doors, internal doors, even garden gates).

It should be mentioned that some self-contained window/door alarm systems also rely on the use of magnetic reed switches. The only difference here is that the switch section is fully integrated with the control unit. Some of these devices can even double as a door chime when disarmed. There are also wireless reed switches for those who want to avoid wiring, although these are less common.

Magnetic reed switches are generally located in a separate loop, because they will only sound the alarm once if the intruder leaves the door or window open after entering. If they are positioned in a loop together with other sensors, this loop can easily be nullified by leaving the reed switch and its magnet well apart from each other. No other sensor can sound the alarm after it has been sounded by the reed switch.

Wire contact systems are old-fashioned but can still be found occasionally. The wire contact switch relies on a wire or thread connected both to the switch and to another fixed point. The switch will sound the alarm if the thread is put under tension or slackened. Since the thread generally is an electrically conductive wire, the wire also functions as a sabotage alarm: the alarm is sounded if the wire is cut or broken. Wire systems of this kind are generally used for protecting windows and walls (Figure 26) or as an external alarm in a garden.

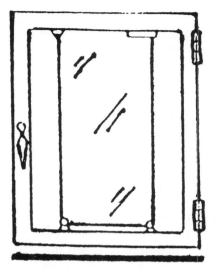

FIGURE 26
Wire contacts

WINDOW FOIL

Adhesive window foil, although now clearly dated as an alarm sensor, is still in common use, especially in shops. It is not used often in residential alarm systems. Window foil looks like silvery lead foil and can be found around the edges of a window (Figure 27). The foil consists of thin metallic strips made of a lead-aluminum alloy that are cemented to the protected surface, usually glass in a door or window. The metallic strips are connected to a closed electrical circuit. If the protected material is broken, and the foil as well, the circuit opens and initiates the alarm. (Foil is also called tape.)

Window foil has many disadvantages, the main one being that a foil alarm is a one-time alarm only. When the window has been broken and the alarm has sounded once, anybody is free to enter until the window and the circuit have been repaired. Another disadvantage is that the foil deteriorates with age and after a few years often becomes too

FIGURE 27
Window foil

ALARM SENSORS

brittle to function. Yet another disadvantage is that the foil will not sound the alarm if the intruder simply cuts the window open with a glass cutter, without breaking the foil. Then repair or replacement is necessary. Finally, windows protected by foil can generally not be opened. For these reasons other alarm sensors, such as glass-breakage detectors, are better investments and therefore more common.

If the foil is installed on a window that occasionally needs to be opened, it will be connected to a contact strip, which is used to disconnect the wires from the window foil block. The spring section is mounted on the window itself, while the contact plate is on the window sill. The metal tabs on both sides of the switch must make good contact.

Under normal circumstances, the connection between the window foil and the control unit is active only when the window is closed. Window foil is used with the NC circuits of the system. Therefore, it must be disconnected if the window is to be opened. If so, the switch must be prepared for a coiled jumper wire that will temporarily bridge the gap between the two parts of the contact strip.

There have been attempts to make more reliable versions of window foil. Certain manufacturers of insulated windows include a very narrow and thin metal strip in the glass, usually hidden by the rubber strip used for insulating the window. They reason that the strip will break if the window is broken. This alarm system is safer: the metal strip is effectively invisible and securely located because it is protected by the glass. However, this is also a one-time-only alarm and suffers from most of the same disadvantages as ordinary window foil.

WINDOWPANE-MOUNTED GLASS-BREAKAGE DETECTORS

Glass-breakage detectors (Figure 28) come in several different types, although they all are designed to detect when a window is broken or otherwise removed from its frame. In either case, the sensor is attached to the glass on the inside of

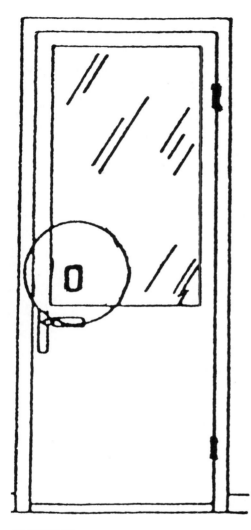

FIGURE 28
Glass-breakage detector

the window. The sensor is fastened with double-sided adhesive tape but can be easily removed if so required. There are versions available both as NC and NO switches. One detector per window is generally adequate unless the window is very large, such as in major offices or in shops.

The three most common types of glass-breakage detectors are the mercury-switch type, the weighted-arm type, and the electronic or "tuning fork" type.

The mercury-switch type relies, obviously, on mercury. At room temperature, mercury is a liquid metal that conducts electricity. The detector is designed to have a small pool of mercury near an NO switch. If the glass in the window is broken or severely shaken, the switch will be tilted or moved, and the mercury will come into contact with the switch terminals. This closes the switch and triggers the alarm.

The weighted-arm type uses a movable internal contact that is weighted so that it remains in one position. However, most vibrations will make the arm move, which will either complete or break an internal switch contact, depending on how it is designed. The contact triggers the alarm.

The electronic glass-breakage detector uses an internal tuning fork that vibrates when the window is broken or jimmied. The alarm is triggered by the vibrating tuning fork.

In the same way as window foil, glass-breakage detectors of these types can use a contact strip if the window sometimes needs to be opened.

All these glass-breakage detectors share the same problem: they are susceptible to any vibrations in the window, including those produced by the wind. So, there may be numerous false alarms, especially if the window is not fixed tightly in its frame.

Therefore, other types of glass-breakage detectors have been introduced. They work by detecting the sound frequencies emitted by breaking float glass or shattered window frames. This detector can detect and identify the special noise that is heard when glass is smashed or a diamond cutter is used or when metal hits glass. For this purpose, the detector incorporates a piezoelectric microphone with a resonating frequency within the range emitted by glass being broken. Another option is a microphone sensitive to frequencies in that range, above 60 kHz. Consequently, this glass-breakage detector is not sensitive to lower frequencies, such as from heavy traffic.

The detector is fixed to the glass it protects, although some types can also be put on adjacent walls. The latter types are detailed below.

The type of glass is important because the majority of these glass-breakage detectors react only to the sound of breaking float glass, not to the sounds of breaking laminated, tempered, or wired glass. Some of these detectors are also susceptible to false alarms and can be set off by the sound frequencies of rattling keys or bottles.

VIBRATION DETECTORS AND INERTIA SENSORS

A vibration detector is a sensor placed on a wall or window frame to register vibration caused by blows, drilling, or breaking glass (Figure 29). Self-contained units, affixed to doors, are also available. The sensor will signal an alarm whenever it registers the kind of vibrations it is programmed to identify as intrusive.

Vibration detectors, or shock sensors, come in different types. Some rely on mechanical means, while others work on electronic principles.

The most common vibration detector is still the pendulum alarm, a mechanical detector that relies on a pendulum switch. A pendulum switch is a switch used to sense vibration or motion, and is designed with a set of NO contacts that touch

FIGURE 29
Vibration detector

each other when the switch is moved or shaken. The switch can also be NC, with the pendulum breaking the circuit when moved or shaken.

Even though the sensitivity of these older sensors can be adjusted manually by a screw, they have a tendency to give false alarms: low-frequency noise, such as from heavy traffic, creates enough vibration to set off the alarm. So, a new type of vibration detector has been introduced, the piezoelectric sensor, which, in appearance, is similar to the older types. This sensor is often used on walls, windows, and doors.

This sensor responds only to high-frequency vibrations, such as those emitted by breaking objects. The sensor contains piezoelectric crystals, a crystalline material that will develop a voltage when subjected to mechanical stress or severe vibration. The developed voltage triggers the alarm. The sensitivity can be adjusted, and the chosen sensitivity will be remembered by a memory circuit. Since semiconductors are used, the piezoelectric sensor has no moving parts. In fact, the piezoelectric sensor is a kind of glass-breakage detector but is not limited to being affixed to the window pane.

Similar devices work according to a different principle: these are acoustical detectors mounted near windows, generally in the ceiling. The detectors are able to guard a number of window panes within its range.

This detector is sensitive to the noise of breaking glass, shattering wood, and other sounds signifying a break-in; its range is often around 15 meters. The detector contains an advanced digital filter, ensuring that lower frequency background noise will not trigger the detector. The filter usually sets the lower limit at 5 kHz, but this is often adjustable to between 2 and 10 kHz. Likewise, noise that builds up gradually—such as from cars, aircraft, and vehicle brakes—will not trigger the alarm. To a great degree, this prevents false alarms.

The system works on the principle of audio discrimination, the process of electronically separating normal everyday sounds (e.g., voices, telephones) from break-in types of noises (e.g., breaking glass, prying metal, or forcing a door open).

Highly sophisticated sound detectors can easily respond to the sound of a window's being smashed. The detector responds to the brief time lapse between the sound of the window's being smashed and the tinkle of falling shattered glass. Some of these detectors are so sophisticated that they can even distinguish between the different sounds of a window's and a bottle's breaking.

There might also be a monitoring function that allows a central security station to listen in on what is happening in the room through the use of an automatic dialer or some similar piece of equipment. This listen-in function is made possible by the electret microphone built into the sensor. These alarm sensors generally, but not always, work with NC switches.

The most recent type of vibration detector is the inertia sensor, which looks virtually identical to the other types of vibration detectors. The inertia sensor is an intelligent vibration detector; consequently, it requires a special control unit to analyze the complex signals from the sensor. The principle of this sensor is in a comparatively heavy contact element that rests on a contact surface, which is mechanically connected to the cover of the detector. The contact element doesn't vibrate with the frequencies that occur when glass is broken or when metal strikes glass, but the rest of the sensor does vibrate.

Higher frequencies give rise to a burst of short-duration breaks that are analyzed in the control unit. When the number of short-duration breaks reach and exceed a threshold value, the alarm is triggered. This happens only when the vibrations are strong enough to indicate a real break-in attempt. Clearly, this process of determining the likelihood of a real intrusion is very complicated, and only a comparatively complex control unit can do so. These sensors have control units in addition to the central control unit of the alarm system.

Inertia sensors are also often found in cars, in trailers, and on containers. One such sensor, with its special control unit, is enough to guard the entire perimeter (i.e., the shell of the construction). When properly installed, the inertia sensor is a very reliable alarm sensor.

INFRASOUND DETECTORS

This is a more modern type of detector, sometimes used for perimeter alarms in houses and in other relatively enclosed objects, with a floor space not larger than 400 square meters. One detector is adequate even in a house with several floors, as long as the total area is not too big. It works by detecting sound within the frequency range of 1 to 4 Hz. This is called infrasound because it is below the level normally audible to the human ear. The change in air pressure that takes place when a window or a door is opened produces such sound. Changes in air pressure always take place when the enclosing material changes its nature.

It is perfectly possible to move around in a building when the alarm system is armed, as long as no door or window is opened. Because only one sensor is necessary, this alarm is very quick and easy to install.

The disadvantage of this system is that it is a one-time-only alarm. If an intruder leaves a door or window open, the alarm sounds only once, when he actually opens the door or window. Therefore, infrasound detectors are usually used with other types of trap alarms.

The infrasound detector can be hidden almost anywhere (e.g., under a staircase, behind a cabinet or curtain). A good sensor of this type automatically compensates for natural infrasounds emanating from strong winds or some other source.

FIELD-EFFECT SENSORS

The field-effect sensor relies on the principle of electrical capacity changes between the guarded object and earth or between extended conductors or foils. In effect, the sensor works similarly to a radio transmitter and receiver.

If somebody is moving in the field between the transmitter and the receiver, there is interference in the received signal. This triggers the alarm. For this reason, these sensors are also known as field-disturbance sensors.

Capacity changes can be indicated in different ways. If a metal safe is to be protected, it can be connected to a frequency-determining resonance circuit. This circuit controls an oscillator, whose output frequency will be compared to a fixed frequency within the range of 100 kHz to 10 MHz. The variations in frequency are detected and analyzed, and if a predetermined threshold value is exceeded, the alarm will sound.

Likewise, a building can be fitted with two encircling conductors, approximately 1 meter apart. One conductor will be the transmitting antenna (in effect, the antenna to a long-wave radio transmitter), while the other will be the receiving antenna (Figure 30). The system detects and analyzes the differences in the received signal that are caused by an intruder's approaching the protected area.

Field-effect sensors generally consist of two or more wires running parallel with the protected area, with the wires connected to the wall or fence by insulators. Particular care should be taken to position them far away from other metal objects, since such objects may limit the range of the detector. Walls, internal walls, corridors, and other locations

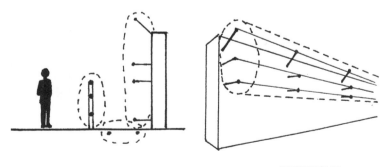

FIGURE 30
Field-effect sensors

can be protected in this way. In a corridor, for instance, the two wires will run one on each side; the wires might be up to 300 meters long.

Alarm systems of this kind are already in many places, but they will definitely be more common in the future.

Another variant is the capacitance detector. Such a detector often has a metal plate on which the protected object is positioned. If an intruder approaches too close to the object and the metal plate, the electrical capacitance in the plate will change and trigger the alarm.

SOUND DETECTORS AND HEAT DETECTORS

Free-standing portable sound detectors can be placed almost anywhere and are extremely easy to activate. A suspicious individual will put one on a table before he goes to bed. The sensor then listens for intruders in the room where it is located—and possibly in adjacent rooms, as long as the internal doors are left open.

Sound detectors are also very prone to false alarms. The problem is basically that they are too good. They can be triggered by perfectly ordinary noises outside the protected area (e.g., in the street).

The most common type of sound detector is the previously described glass-breakage detector.

Sound detectors are often mounted on safes, to sound the alarm if somebody attempts to crack the safe. This detector consists of one or more microphones connected to an amplifier programmed to respond only to the type of noise expected that signifies a break-in. This makes certain that the occasional noise of a broom or a vacuum cleaner hitting the safe during cleaning does not trigger the alarm.

The microphones are attached to the object with a strong magnet, and a plunger switch or magnetic reed switch sabotage alarm is included, so the alarm will be sounded if somebody attempts to remove the detector from the object.

Sound detectors are often mounted on the walls of bank

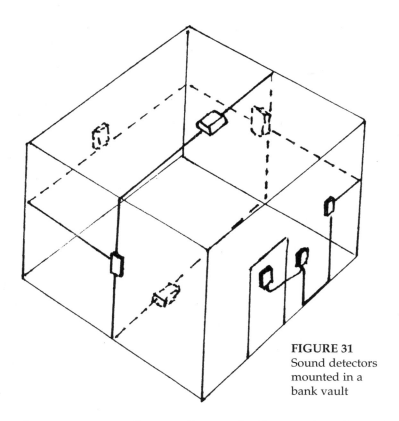

FIGURE 31
Sound detectors mounted in a bank vault

vaults—on every wall, as well as in the floor and in the ceiling (Figure 31).

A sound detector is often mounted together with a heat detector. (The latter device should not be confused with the infrared detector.) The heat detector will only register the heat of a cutting torch or fusing burner.

PRESSURE MATS, PLUNGER SWITCHES, AND CONTACT STRIPS

Pressure mats look like rubber floormats and are commonly hidden under the carpet, especially under fitted (e.g., wall-to-wall carpets and linoleum) at strategic places in the

ALARM SENSORS

building. Often such points are in front of certain doors and windows, at the foot of the stairs, or directly in front of a safe or important object. It is also common to use a series of pressure mats in a staircase, so that it is virtually impossible for an intruder not to step on at least one of them if he wants to get to the next floor. Because staircases are often of different sizes, the pressure mat also comes in several different sizes (Figure 32).

The pressure mat is a large NO switch. The mat contains two grids of switch contacts separated from each other by a nonconductive material, which has been perforated several times, so the contacts can be brought together through the perforations. The contacts are pressed together only when a person of sufficient weight steps on it. (Generally, it requires an adult; children and pets should not trigger the alarm. However, certain inferior brands of pressure mats may be triggered by cats or dogs.) Then the NO switch is closed, and the alarm goes off.

If the pressure mat is hidden in an unsatisfactory way, the

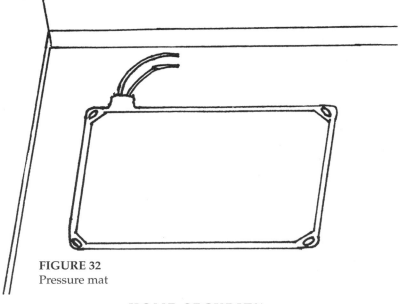

FIGURE 32
Pressure mat

outline of it might start to show through the carpet after long use. The intruder can then easily avoid it.

A plunger switch is a mechanical device located on doors and windows to detect entry or tampering (Figure 33). In fact, it is an ordinary spring-loaded momentary switch designed for use in alarm systems, having a depressible plunger or button. The plunger in itself is oversized, while the body of the switch is designed for mounting on doors, windows, or control units. The plunger switch is easy to bypass, as long as it is known to exist. An intruder will simply keep the plunger under pressure (e.g., with a piece of celluloid), so that the device does not set off an alarm, while he removes the switch from the protected area.

The plunger switch is often used in control units as a way to prevent tampering with the system. In this application, switches are mounted on the front or back of the control unit. The alarm is sounded if the door is opened or the control unit is removed from the wall, where it is more difficult to remove the switch because of its location.

In the same way, plunger switches can be used to detect when a door or window is opened. In this case, the switch is mounted so that the plunger will be

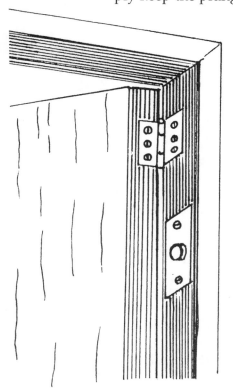

FIGURE 33
Plunger switch

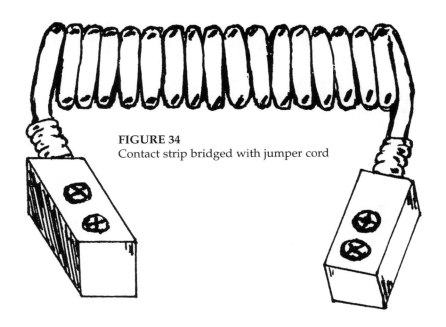

FIGURE 34
Contact strip bridged with jumper cord

depressed when the door or window is closed. The switch is then mounted on the hinge side, in the door jamb, and is virtually impossible to see because it is exposed neither from the inside nor the outside when the door is closed. This type of switch is, of course, an NC switch. Whenever the door is opened, the alarm sounds.

A contact strip mechanism is an open switch, sometimes used on windows and doors, usually together with window foil and glass-breakage detectors. The contact strip is similar in shape to the reed switch and is also composed of two parts. Here, too, one part is mounted on the window or door, while the other is mounted on the frame (Figure 34).

The advantage of the contact strip is that the two pieces can be bridged with a jumper cord. This is highly useful when a window or door (e.g., a garage door) must be left open but the alarm system is still armed.

IONIZATION DETECTORS

Ionization detectors are not yet, as far as it is known, in use anywhere, but this detector might well be the alarm sensor of the future.

An ionization detector would work according to the principle of the Kirlian effect, which is that living matter is surrounded by a "force field" that ionizes the surrounding atmosphere. This produces an aura around the living being when photographed in fields of electrical current.

The Kirlian effect is not universally accepted because the reason for this phenomenon is so far unknown. (The phenomenon has also been frequently abused by sensation-seeking parapsychologists.) However, this has not prevented research in ionization detectors.

An ionization detector functions in the following way. It is well known that if the atmosphere is ionized, the electrical conductivity of the atmosphere is changed. An ionization detector registers the atmospheric conductivity. Any changes caused by the ionization of the surrounding atmosphere by the force field around a human being will be detected.

In the Soviet Union, the Kirlian effect was accepted as valid for further research. The demise of the Soviet Union has, somewhat unfairly, discredited much of the research that went on there. Nonetheless, it is probably only a matter of time until sensors of this type will be in production.

PHOTOELECTRIC CELLS AND INVISIBLE BEAM DETECTORS

Photoelectric cells have been in common use for years, especially in shops, for detecting persons walking into a room or through a door. In old-fashioned systems of this type, visible light is used, and the person entering has no trouble at all noticing and evading the photoelectric cells, should he choose to do so. Currently, invisible infrared light is used instead of visible light (Figure 35).

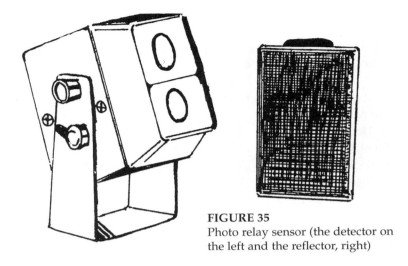

FIGURE 35
Photo relay sensor (the detector on the left and the reflector, right)

The modern invisible-beam detector, or photo relay sensor, is designed to project a narrow infrared (IR) beam across the area to be protected. For this reason, sensors of this type are also sometimes called active IR detectors. The beam is aimed onto a small reflector, which reflects the IR beam back to the invisible-beam detector, which has a built-in photoelectric eye that is sensitive to IR light. This eye constantly monitors the area, and as soon as somebody interrupts the beam between its projector and receiver, either completely or nearly so, the alarm goes off.

Most invisible-beam detectors of this kind are able to protect an open area up to about 10 meters wide. Larger units are conceivably capable of a range of several hundred meters. These units are commonly used outside buildings, such as in private gardens or parks (Figure 36). Rooftops are also sometimes protected by these detectors. With the exception of really sensitive installations, these detectors generally trigger security lights rather than sirens, since the chance of false alarm is very high.

FIGURE 36
IR beams arranged for perimeter protection

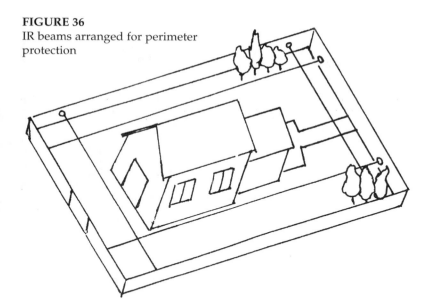

The most common installation is with two invisible parallel beams. A small animal such as a bird will then not trigger the alarm, but a human-sized intruder will break both beams and trigger it. Likewise, mist or falling snow, rain, or leaves will not trigger the alarm. The range is generally up to 150 meters but is shorter in cold climate regions. Many detectors of this type can function even if covered by snow or frost, but they are more susceptible to dirt and may need frequent cleaning in certain environments.

The emitted IR light is modulated so that the receiver can identify it without being disturbed by other sources of IR light, such as sunlight. The IR beam might also be reflected by way of reflectors (Figure 37), which increases the range.

The invisible light used in the beam consists of IR light, so these detectors are easily found and evaded (for instance, by using IR goggles during the break-in). An intruder can also locate the beams by blowing cigarette smoke at them. A combination of IR goggles and cigarette smoke often produces the best result.

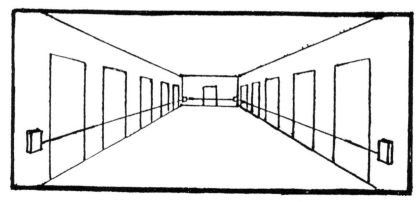

FIGURE 37
The use of IR reflectors to protect a corridor

PASSIVE INFRARED DETECTORS

A passive IR detector (or PIR, as it is commonly known in Britain) receives and measures IR energy from other objects. In effect, it is a heat detector, although of a very different type than those discussed earlier in this chapter.

The passive IR detector is usually mounted high on the wall, either in a hall or a large room that must be traversed to reach the sensitive areas. The passive IR detector works by picking up an intruder's body heat within the detection range and triggering the alarm. The detector is extremely sensitive and will pick up even the smallest variation in temperature, both above and below the normal.

The passive IR detector works through a built-in pyroelectric sensor. Simply speaking, this sensor is able to indicate heat by producing an electric charge. The pyroelectric sensing material is polarized by IR radiation, producing a voltage proportional to the rate of change of incident energy. The pyroelectric sensor monitors the room electronically protected by the alarm; it is surrounded by a reflective surface that collects IR energy. Because the sensor is sensitive to heat (i.e., IR energy), it senses any change in the level of IR energy, including the presence of an intruder radiating heat. This abrupt change will be readily detected, triggering the alarm. A resting human radiates approximately 100 W, so we are all major sources of IR radiation. However, mere radiation is not enough to trigger a passive IR detector; the heat source must also be moving through the zone monitored by the detector.

The passive IR detector divides the area to be monitored into zones and segments (Figure 38), typically up to 7 meters across (usually 76 degrees wide, but this can vary), 12 meters deep, and around 70 centimeters high. Only these zones and segments are monitored, and intruders above or below the

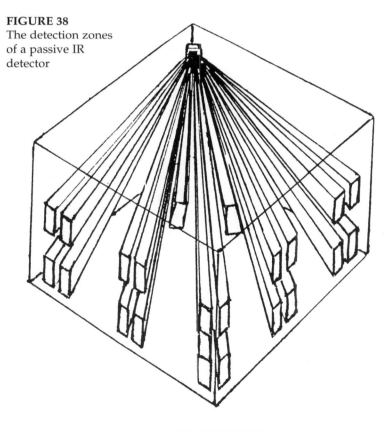

FIGURE 38
The detection zones of a passive IR detector

zones will not be detected. Therefore, the adjustment of the detector during installation is very carefully done to ensure maximum protection as well as to prevent false alarms caused by pets or small children. For these reasons, the alarm is generally positioned at the level of the head of a man-sized intruder or higher, at least 1 1/2 meters above the floor. The detector is most often positioned along a wall or in a corner, so that it can have a free line of sight to as much of the room as possible, including the front door if at all possible.

A passive IR detector typically looks at a zone of six or more separate segments with unwatched "aisles" between the segments. A separate-zone design gives the unit wide-angle coverage of a room. Because the background radiation of each zone is slightly different, it is virtually impossible for an intruder to match them all.

Self-contained IR sensors generally cover a wider area and have more zones, but typically protect an area up to only 8 meters deep. There is also a lower zone in each segment, usually ensuring that an intruder is unable to slip under the monitored zone. Sometimes, especially if pets are on the premises, the lower zone will be turned off.

There are also IR detectors on the market that can monitor areas of up to 50 meters in length. Some of these detectors can have their detection range varied by changing the lenses. However, the IR detector is prone to false alarms, being triggered not only by such obvious problem sources as cats and dogs, but also by small pests or strong sunlight through a nearby window. Even being too near any quick-changing source of heat or cold (e.g., central heating producing radiated heat and warm drafts, an air-conditioning duct) might produce a false alarm.

When an IR detector is switched on, it first balances itself based on the amount of IR radiation given off by the various sources in a room, such as background radiation from walls, furniture, and floors. If an intruder later enters the detection zone, he alters the amount of radiation received by the detector, which results in an alarm. That the IR detection system first has to balance itself means that if the power has been switched off for some time, the sensor will require approximately 90 seconds to be operational. This is not necessary when the system has been merely disarmed while retaining power.

The detector will not respond to slow variations in background radiation, because amplifying circuits limit the detected variations to a predetermined range of possible speeds of the IR source before the circuits allow the detector to sound the alarm.

In advanced units of this type, a threshold circuit also ensures that a signal is large enough to represent an intruder. In this case, the alarm will not be set off by a pet if it is not close to the detector. The positioning of the detector and its segments of protection also determine the sensor's sensitivity to false alarms.

Every IR sensor is most effective in detecting movement across the segments rather than toward or away from it (Figure 39). There have been some lab tests regarding the possibility of approaching the sensor directly, very slowly so that it will not detect a sharp increase in temperature and, consequently, not sound the alarm. In some of these tests, a panel of thick glass was used by an "intruder." In theory, an IR detector can be foiled by the glass panel put in front of it: slowly moving the panel in from the side seems to be most effective against the detector's picking up the presence

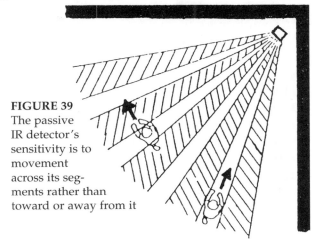

FIGURE 39
The passive IR detector's sensitivity is to movement across its segments rather than toward or away from it

ALARM SENSORS

of an intruder. However, with extreme care, an intruder could conceivably stay behind the glass panel and approach the IR detector directly from the front, the detector supposedly being blinded by the thick glass. (Cardboard would also be reasonably effective because it too diffuses the heat signature.) However, such experiments have generally not been very successful indoors: even slow movement causes some warm air turbulence that the sensor is able to detect. The glass panel trick often works outside a building, but there an IR detector would be very prone to false alarms in any case and not much relied upon.

An IR detector can also be mounted in the ceiling. If mounted as high as 6 meters, the protected area can be as large as 6 x 20 meters, and all movement will naturally occur across the segments, thus facilitating detection.

Passive IR detectors, although generally used to monitor movement in rooms or along corridors, can also be used vertically to create an alarm curtain, for instance in front of a wall of paintings. Naturally, the detectors can also be used to create a horizontal curtain to detect a break-in through the roof, ceiling, or floor.

Most passive IR detectors have built-in tamper switches to prevent somebody from removing the unit from the wall or attempting to disarm it.

Infrared sensor systems, whether self-contained or part of a separate-component system, are easy to place. The sensor is generally positioned so that its zones of detection include those points expected to be the entry points of an intruder, such as the door and windows. The intruder should be forced by the layout of the building to walk into these zones. Furthermore, the sensor should be placed high, at least 1 1/2 meters above the floor. This is necessary if the lower portions of the zone are to be used.

The sensor unit should also be positioned so that its field of view does not include solar-heated walls, direct sunlight, or other sources of variable heat such as heaters, air conditioners, lamps, or any other object that might change its temperature quickly. Older IR sensors might give false alarms under these circumstances. Contemporary models will not as readily give false alarms, but their detection ability is somewhat impaired. In today's models, every segment is divided into two channels. The detector sounds the alarm only if the change in temperature is different in both channels; this avoids sounding the alarm when, for instance, a radiator heats up. An intruder, however, will disturb first one channel and then the next, thus triggering the alarm.

Self-contained IR sensor systems incorporate entry and exit delays, typically 15 to 20 seconds in length. Otherwise, a remote key switch might be used. Backup batteries, as well as external warning devices, can generally be connected.

There are also smaller self-contained IR detectors, which generally work with only one coverage zone and are supposed to guard only small areas near doors, windows, trailers, and suchlike. If a heat source is detected, the built-in siren will sound for about one minute, after which the unit will be reset. An entrance and exit delay of a few seconds is generally built into these units. Power is typically provided by a standard 9-volt battery inside the detector unit.

It is sometimes possible to hide the IR detector so that an intruder cannot find it. Certain types of plastic foil exist that will not let visual light through yet remain transparent to IR (heat) radiation. Because this foil is available in various colors, you can hide the IR detector in a wall or picture frame and then cover it with foil to make it appear like a work of modern art.

Contemporary IR detectors are often very small, some being no larger than a matchbox.

MICROWAVE MOTION DETECTORS

A microwave motion detector is a trap alarm with a single-unit transmitter/receiver that reacts to distortions in the timing of its return signal. This device works in effect like a small radar system, using an electromagnetic field composed of high-frequency radio waves generated over the protected area. Most detectors of this type have a range of 30 meters or less.

The microwave motion detector relies on microwaves, very high-frequency radio signals in the frequency range around 9 GHz. These radio waves form a pear-shaped lobe (Figure 40) that will detect any intruder. As these microwaves are reflected off solid objects, they reveal the presence of an intruder entering a protected area, because his movements will cause a disturbance in the reflected radio waves (i.e., the radiated radio frequency [RF] field) sensed by the device.

This disturbance is a modulation of the field referred to as the Doppler effect. The Doppler effect is the difference in frequency of one (original) wave to another (reflected) wave that is superimposed on the field. A frequency shift occurs when a signal source and a receiver are moved relative to each other. According to the Doppler effect, the reflected signal is of a lower frequency than the emitted signal if a reflecting object, or a human, is moving away from the detector. Likewise, the reflected signal is of a higher frequency than the emitted signal if a human is approaching the detector. Because the signal moves at a constant rate, the return trip of the reflected signal should take the

FIGURE 40
Microwave motion detector and its detection zone

same length of time as the emitted signal's outward trip. However, an intruder who moves into the path of the signal distorts the timing of the return signal, thus producing the Doppler phenomenon. In either case, this frequency shift is sensed by the microwave detector unit's receiver. Whenever it detects such a change in the reflected signal, the unit will trigger the alarm.

The frequency of the microwaves is very high, so the wave length is very short. This means that even a very small movement will result in a large frequency shift. However, the detector is less sensitive to objects moving across its field of detection (Figure 41) than it is to objects moving toward or away from it.

Most microwave detectors also demand that a person move with a certain speed to be identified as an intruder. These detectors have the ability to reject faster-than-walking speed. This prevents numerous false alarms, since many microwave motion detectors, especially the older models, work too well. Because these detectors' signals can penetrate beyond the walls of a house, they might identify, for instance, a passing truck as an intruder. External radio sources can also trip these units; the RF energy radiated by many electric devices will disturb the functioning of the unit. Such electric devices include radio transmitters, especially cit-

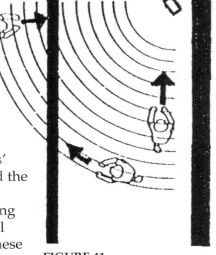

FIGURE 41
The microwave motion detector's greatest sensitivity to movement toward or away from it, with sensitivity to movement in another room

izen band radios, but also cable television systems, motors, transformers, and even neon signs and fluorescent lamps. When the lamps are switched on, the gas in them ionizes to become a fluctuating reflector, which can easily trip an alarm.

A false alarm might also be caused by cooling fans or even the roof of a warehouse if the roof rises and falls in the wind. False alarms can also be caused by the movement of water in pipes: the sensor has the ability to see through glass and thin walls of plaster or plastic. (A microwave motion detector is an extremely small radar, so the microwaves are completely harmless, unlike the radiated energy from a large radar, such as that used for air traffic control or military purposes.)

There are also combined microwave–passive IR detectors, which are very reliable because these two sensor types complement each other. The microwave detector determines that the object moves within a certain speed limit, while the IR detector checks whether the object emits heat or not. These combined detectors come in different types, with ranges of up to 65 meters. Advanced models are even immune to false alarms from pets because they use special imaging technology to distinguish between the shape and size of humans and animals.

ULTRASONIC MOTION DETECTORS

An ultrasonic movement detector is a single-unit transmitter/receiver that signals an alarm when its steady pattern of inaudible sound waves is disturbed. This detector works in a similar fashion to the microwave sensor, but it uses ultrasonic sound waves instead of microwaves, and the principle is still that of radar.

The ultrasonic detector works by continuously transmitting ultrasonic sound waves of such high frequency, around 40 kHz, that a human cannot hear them. The typical upper limit of human hearing is 20 kHz.

The sound waves bounce off the hard surfaces in a room, in effect producing an echo, and thus reflected are picked up again by the same unit. An intruder entering the protected room will interfere with the frequency of the reflected sound: because of the Doppler effect, the reflected sound waves will experience a frequency shift if they are reflected from such an intruder. When the intruder moves within the field, the unit's receiver detects a change in the reflected signal and triggers the alarm.

Because of the nature of sound, the ultrasonic detector produces a teardrop-shaped "cone" of ultrasonic energy, similar in shape to what is illustrated in Figure 40. This cone will expand horizontally and vertically away from the unit across a relatively broad area, typically up to 9 or 10 meters in a forward direction and 7 meters wide. However, the shape and size of the cone varies with the acoustical characteristics and shape of the room in which the sensor is located. Walls and glass windows will be penetrated only to a minor extent. Furniture will not be penetrated, and this may create so-called dead zones in which the detector cannot function. So, even something as minor as moving some furniture can totally change the protected and dead zones in a room. You must then retest the detector and perhaps move it to get the coverage you desire. Even a minor move of 10 to 20 centimeters may work wonders.

The fact that the ultrasonic cone cannot penetrate walls enables a knowledgeable intruder to blind the sensor. An ultrasonic motion detector (unlike a powerful microwave motion detector) will be foiled by a thick sheet of paper or cardboard positioned in front of the detector, similar to the way that an IR detector can be foiled by a thick sheet of glass. It is safest to move the paper sheet slowly in front of the detector from the side (car thieves can often do this easily by opening a side window). But with extreme care, the intruder can also hide behind the cardboard and slowly approach from the front. This blinds the detector because it is unable to broadcast the ultrasonic signal that the system can receive and analyze.

Another technically more complicated but safer way to

bypass many (but not all) ultrasonic motion detectors is to jam the ultrasonic receiver. But only the most sophisticated burglar would attempt this. He would first determine the frequency of the ultrasonic transmitter, either from technical data available in the manufacturer's catalog or by purchasing a specimen of the same model for testing purposes. Alternatively, he could easily determine the frequency on site—before he enters the cone that will sound the alarm—with the help of an ultrasonic microphone, amplifier, and battery-powered frequency scanner. He would then use a battery-powered ultrasonic transmitter of his own from a safe distance to broadcast an ultrasonic signal of the right frequency, which will slowly rise until the security system's ultrasonic receiver is so overwhelmed that it no longer is capable of analyzing the difference in frequency that it needs to identify a moving object. At this point, the burglar can move his transmitter close to the detector, after which he is able to move around freely.

There are also ultrasonic detectors that work with separate transmitters and receivers, and they often have a longer range.

The detector is most sensitive to movements directly toward or away from the sensor (Figure 42). This is unlike the passive IR detector, which is most sensitive to movements across its protected area, but similar to the microwave motion detector. The reason is that both these detector types rely on the Doppler effect. Ultrasonic detectors, therefore,

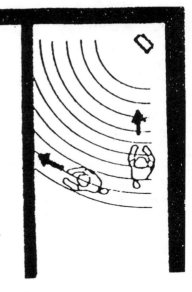

FIGURE 42
The ultrasonic motion detector: also most sensitive to movement directly toward or away from the sensor

are more commonly used for protecting doors and hallways than passive IR detectors.

Ultrasonic detectors are usually located in the corner of the protected room, fixed to the wall or aimed down a hallway or at an entrance, such as the front door.

These detectors are fairly prone to false alarms produced by loud noises (e.g., ringing telephones) or moving objects (e.g., draperies near vents or fans, air currents, drafts of moving air from heating or cooling vents). For this reason, most ultrasonic detectors include a control for adjusting the sensitivity of the device. This ultrasonic detector cannot be used when pets are present, because there are no unprotected areas within the coverage zone. The use of ultrasonic detectors seems to be decreasing because of the number of false alarms experienced by most units of this type.

Many ultrasonic detectors have a delayed alarm, which makes the first steps into a protected area go apparently undetected until it is already too late.

VISIBLE-LIGHT DETECTORS

Very simple in design, visible-light detectors can only be used in closed locations where no natural light can enter. Their main use is with other detectors in a bank vault, which is supposed to stay dark when locked. The detector simply senses whether the level of visible light in the location is rising. If so, the detector sounds the alarm.

VIDEO DETECTORS

A video detector is constantly monitoring the object that it guards by merely "watching" it. The detector, a modified video camera, checks the level of black (the number of black dots, as opposed to the number of white dots, in a picture) in certain interesting areas of the video input. If this level is changing slowly, the detector will not trigger the alarm, because this change indicates some natural occurrence (e.g., sunset).

However, a drastic and immediate change will trigger the alarm because this means that a person or an object has entered the field of vision and is now near the protected object.

Video detectors are reliable but fairly expensive. For this reason they are not yet in common use. They will probably be more common in the future, especially since it is also possible to transmit the video signal by radio or the telephone network with the help of a modem. Then a remote-control station can view and interpret the video.

The latest detectors of this type are even more advanced than the standard video detector. A number of cameras can be used to constantly monitor the protected zone, and the video output from each camera can then be processed by an intelligent image processor based on neural network technology. This will identify any intruder automatically.

The system works by initially learning the characteristics of the natural state of the environment under observation, including such moving items as shadows, branches of trees, or level-crossing barriers. This initial learning state does not need to last for more than a minute, and the system can even be taught to ignore certain dynamic events, such as guard patrols or vehicular movement.

When the system enters operational mode, it identifies intruders in the field of view by recognizing abnormal patterns of movement. The system immediately sounds the alarm and automatically trains a high-resolution camera on the target for identification and video-recording purposes.

This advanced system is not yet in general use but can be found at certain British defense research installations. It can also be configured to be portable. Expect it to be common in the future in most high-risk installations and eventually also in private residences. The latest video cameras are small enough to be hidden in ordinary jacket buttons.

BARRIER SENSORS AND ANALYZERS

A perimeter barrier is a wall, fence, or gate marking the

perimeter of the property. Guarding such a perimeter, especially if long distances are involved, requires special sensors and sometimes special control units, or analyzers.

Inertia Sensors

Today, the most popular barrier sensor is the inertia sensor, even though there are various other types of barrier sensors on the market. The inertia sensor, used with a special barrier analyzer, can be easily adapted for use on any fence.

The inertia barrier sensor system is an electromechanical system that relies on special wires (Figure 43). Every wire is connected to a special self-adjusting sensor installed in a sensor post or pole. The sensor is positioned between two horizontal wires that run between the sensor posts. The sensor immediately notices any attempt to spread the wires (for entering between them), climb on them, cut them, or otherwise remove them.

The sensor-connected wires are fitted with vertical springs and self-adjusting connection points, so that they will detect any kind of intrusion. Six wires and sensors will

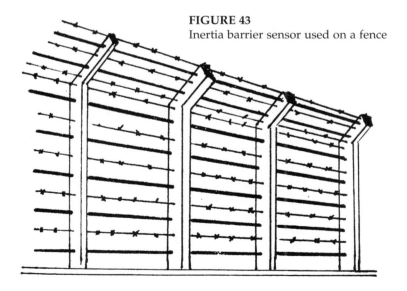

FIGURE 43
Inertia barrier sensor used on a fence

generally be fixed parallel to each other on the posts, so that it is impossible for an intruder to avoid all the sensor wires. This also allows the system to determine the exact height of an intrusion. Another advantage is that a wire group can be disconnected (e.g., during maintenance) without the rest of the sensor post's being affected or impaired.

The sensor posts are generally positioned up to 50 meters apart, although a shorter distance, say 10 meters or so, makes the system able to really pinpoint an attempted intrusion. The posts are made of aluminum or stainless steel, unlike the ordinary posts made of stainless or galvanized steel. Every sensor post is protected individually by tamper-protection technology and monitored individually by the computer in a central security station.

A multiplex computer communications system keeps in touch with every single sensor post and wire through a special information cable that runs along the fence. Every detected intrusion is thus reported by a special unit in each sensor post and monitored and registered by the the central security computer. The system also self-tests continuously, sounding an internal alarm if part of the system breaks down or is subjected to attempted sabotage. Naturally, this self-test also includes the information cable. Every incident will generally be logged into the central security computer if a need arises to verify any events.

Computer communication goes in both directions, however. This means that individual security measures can be adopted in case of an alarm, including activating video cameras and searchlights or arming mines. Such measures can be programmed to be activated automatically, without the need for a manual operator.

As can be imagined, this type of system is very complex and used only in high-risk locations. The most famous system of this type is the Israeli Magal system, which now guards the entire length of Israeli border fences and barriers. An identical system, also produced by Magal, is used by many other countries in military installations and nuclear power plants. The only safe way to circumvent this alarm system is to avoid touching the fence: a really determined intruder will enter only by air.

Military installations such as air bases are often protected by ground surveillance radar. One such system, the Dutch Squire system, has a range of 24 kilometers and can detect a single pedestrian out to nearly 10 kilometers. If you need this kind of surveillance, you'd better be able to afford the price, which can reach higher than $200,000 per radar unit.

Microwave Fence

A common sensor used for protecting a perimeter is the microwave fence. This is, in effect, a radar barrier and works by having a transmitter send a pattern of invisible microwaves to a separate receiver, located up to 300 meters away. This pattern will be distorted by any intruders crossing the path between transmitter and receiver, which then sets off the alarm (Figure 44).

Transmitters and receivers can be positioned along the perimeter of a garden or other open area. The microwaves create a signal field that might be between 2 and 8 meters high and up to 20 meters wide or more.

This sensor is very prone to false alarms because there is no way of distinguishing between an intruder and an ordinary

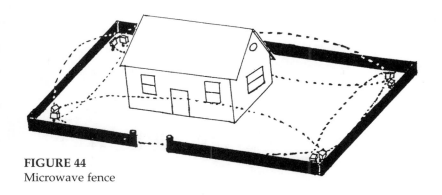

FIGURE 44
Microwave fence

bird or animal. Even moving vegetation might trigger the alarm. On the plus side, the sensor will detect very small changes, so that an intruder crawling slowly into the field will be detected. The sensitivity is adjustable, however, and some systems of this type will indicate only those intruders who are at least walking slowly. The detectable speed can usually be adjusted to between 0.01 and 10 meters per second.

Because the transmitter and receiver can be connected by a synchronizer cable in order to force the receiver to notice only its dedicated transmitter, the system can be sabotaged by cutting this cable—but this will generally trigger the sabotage alarm. The system cannot be taken out by simply transmitting microwaves of the correct frequency to the receiver, because the synchronizer will reveal this as an error. Microwave barriers are generally fenced in to avoid frequent false alarms triggered by the presence of animals.

Geophone Sensor

There is another device used for external alarms, the geophone or seismic sensor (Figure 45). A device that monitors vibrations, it can be installed to detect activity across the ground or vibration caused by the scaling or assaulting of walls and fences. The sensor is, in effect, a vibration sensor. The most advanced models have an effective range of 20 meters.

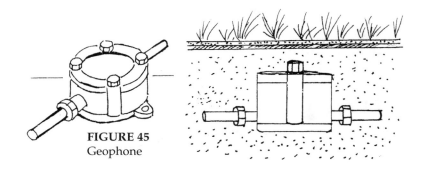

FIGURE 45
Geophone

An older and cheaper barrier alarm system relies on mercury switches as fence alarms. The switches are mounted on the fence and sound the alarm whenever the fence is moved by somebody climbing it or leaning a ladder against it. The mercury switch is also a vibration sensor, although of an older type, and can be circumvented by not touching the fence.

Cable Detector

Yet another system is the cable detector. This consists of special cable capable of changing resistance or capacitance when it is bent or hit. The cable runs through the important parts of the fence and triggers the alarm if the fence is bent to facilitate an entry, climbed on, or cut by an intruder. Any such movement of the fence causes a capacitive change in the cable that, in its turn, triggers the alarm.

Sound Detector

One last system that is fairly reliable relies on a thin tube or hose connected to a microphone. The signals from the microphone are analyzed by an advanced electronic circuit. Any attempts to climb or destroy the fence will be recognized and set off an alarm. The microphone can also be connected to a loudspeaker and monitored in an alarm control station.

All barrier sensors frequently give false alarms caused by animals, strong winds, or some other disturbances that cause the fence to move. The sensors are usually connected to the control unit separately from other types of sensors so that the entire alarm system is not tripped only because of an indication from the barrier sensor.

CHAPTER 4

Personal-Attack Alarms

A deliberately activated alarm, or panic button, is a device designed to raise the alarm when the individual in charge of the alarm feels threatened or is under attack. Panic buttons are always fixed to a certain position, such as near a bed or at a teller's station in a bank.

A personal-attack alarm is used for a similar purpose, but this device is not fixed to a certain location. The personal-attack alarm is carried on the person of the one to be protected and can be activated by its carrier wherever he might be on the premises. Personal-attack alarms are often carried by medical personnel, who regularly have to deal with potentially dangerous patients. These alarms are sometimes carried by the elderly, who fear attack by burglars.

Most professional alarm systems, whether for use by private individuals or corporate or government offices, include one or more panic buttons or personal-attack alarms. Alarm systems in private homes might have two panic buttons, one by the front door and another by the bed. Office alarm systems might have any number of these protective devices.

Regardless of the number of these alarms, they all work in the same way. Panic buttons are wired to a circuit on the alarm system that is always live, whether or not the alarm has been switched on at the control unit. The user needs only to press the button at the first indication that an intruder is trying to enter the premises. The button is designed to be easy to find and press, even in darkness.

Panic buttons used in banks and similar locations are more complicated. This is because most police departments take drastic action as soon as they receive the alarm, such as closing down all public transportation systems in the vicinity. Such responses are very costly and unpopular, and, for obvious reasons, the police do not want to use them unnecessarily. Therefore, panic buttons are designed to be almost impossible to trigger accidentally. They may be either hand- or foot operated and usually must be activated either by pressing with two fingers or by lifting with the foot to minimize the risk of false alarms.

There is also a discreet indicator installed near the panic button, informing the person activating the alarm that it has been activated. This is usually an LED or a similar light that remains lit until the alarm is manually disarmed. There must be no question of whether the alarm has been activated or not. There is also generally a corresponding indicator, coupled with a buzzer, somewhere in the rear of an office, well out of sight and hearing of the front office where the alarm is sounded. The indicator informs a security officer that the alarm has been activated. At the same time that the buzzer sounds, the alarm will go to the local police department through an automated dialing system. In addition to this, it is also common that the security officer will verify the alarm by personally telephoning the police.

A personal-attack alarm is essentially a wireless panic button. This alarm is designed to be carried on one's person, and when activated will trigger the existing alarm system and warning device in the building. Personal-attack alarms rely either on ultrasonic sounds or radio transmissions.

The radio transmitter used as a portable panic button is a fairly obvious design, in effect simply a transmitter capable of activating the alarm system. The transmitter will send a digital code that activates the control unit, and the code will tell who activated the alarm but not the location of the trouble. Of course, this is a disadvantage. However, the plus of a radio system of this type is that the range is generally fairly long: this personal-attack alarm can sometimes be relied upon even outside a house. One way an intruder might try to counter this alarm is by preventing the radio transmission, for instance by enclosing the transmitter in a metal box. Obviously, this would be no easy matter before the alarm is raised.

An ultrasonic personal-attack alarm works in a slightly different way. The ultrasonic signal will be received by one of a number of special ultrasonic receivers, each of which is mounted in every room from which an alarm signal might need to be sent. The ultrasonic signal activates the receiver, which in turn triggers the alarm. An LED on the control unit indicates which receiver triggered the alarm and, consequently, in which room the person who activated it currently is. The identity of the user will not be known, however, if more than one of the personal-attack alarms are in use. Furthermore, the range of this system is much shorter than the range of a radio transmitter.

Both these alarm systems can be connected to a personal paging system. The alarm can then be silent, but (for obvious reasons) this is seldom, if ever, the case. If a silent alarm really is desired, then anybody equipped with a personal paging system can be chosen to receive the alarm.

Personal-attack alarms should not be confused with the so-called personal alarms sold in many retail shops. The latter are small devices that emit a painfully loud, high-pitched screeching noise that will surprise an attacker as well as call for help. But these alarms are not connected to any alarm system; they're only used by people who fear being attacked, especially when they go out. The efficiency of these alarms on a deserted street is not very high, though it might be enough to scare off the occasional mugger.

There are various other kinds of personal alarms. Some, generally the smallest ones, are activated by compressed air and look like aerosol-operated devices. Others are powered by batteries or rechargeable power units. Some of them are activated by pressing a button or squeezing a trigger, but they only sound the alarm as long as the button is pressed or the trigger squeezed. This is especially true of the gas-powered types. More reliable personal alarms work somewhat analogously to a hand grenade: the alarm continues to sound until it is switched off or the power runs out, even if the alarm is dropped to the ground. This is especially helpful if a situation develops into a physical fight.

Some individuals, especially women afraid of attackers of one kind or another, carry a whistle instead of any of the above types of personal alarms. The purpose is the same: to scare off the attacker and summon help.

CHAPTER 5
Fire Alarm Systems

Most household fires start in the kitchen, typically from carelessness or neglect (leaving food or cookware unattended on a burner or in the oven). Most fire fatalities, however, occur in the early morning (between 2:00 and 6:00) and are a result of fires that start in bedrooms and living rooms. Common causes of fire include neglected candles and smoldering cigarettes, portable heaters that have been knocked over, and faulty electrical appliances or wiring. It is obvious that most fires are caused by carelessness. But their result is devastating; a residential home can be totally ablaze in less than five minutes from the start of the fire. The heat can be so intense that just one breath can cause severe lung damage or unconsciousness.

Your first step, even before considering a fire alarm system, should be to acquire at least one fire extinguisher for each floor. Mount the extinguisher on a wall where it is both visible and accessible. Different fire extinguishers are used for different types of fires, so be sure to buy an appropriate type. Don't neglect to read the operating instructions: you, and all other family members, must know what to do in an emergency.

In most cases, it's preferable to be able to clearly distinguish between a fire alarm and a burglar alarm. Not only do you want to call for different professionals in the two different situations, you want to know what danger you're facing, especially if you are inside the building. And if you are outside, you don't want your friendly neighbor to come rushing over with a bucket of water, only to be killed by brutal burglars. Fire alarms sometimes form part of intruder alarm systems, but more often they are separate. The warning sound employed by fire alarm systems is usually distinct from the sound of intruder alarm systems. This is generally true whether the systems are connected or not, except in the simplest alarm systems.

Whether the fire alarm is incorporated in another alarm system or completely self-contained, it will more often than not rely on either one or both out of two radically different types of sensors: the smoke sensor and the thermal sensor. Both types are very common. Yet another type of sensor, the differential detector, is also sometimes used.

The smoke detector (Figure 46) is placed on the ceiling or high up on a wall, because smoke always moves upward. Keep smoke detectors away from drafts created by fans or air ducts, because moving air can blow smoke away from the sensor. Several smoke detectors, either as self-contained units or connected in parallel, are often be seen in bedrooms or at least in a common hallway. The sensor works, as its name implies, by detecting smoke. Remember that most fire fatalities are caused by smoke, not flames. A fire emits poisonous gases that put a sleeping person into a deeper sleep and eventually suffocate him. If you receive no warning, the

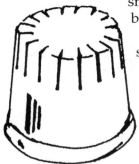

FIGURE 46
Smoke detector

smoke will kill you before you ever become aware of the fire.

There are two different kinds of smoke detectors: ionization and photodiode (or photoelectric). They differ in the way that they detect the smoke that triggers the alarm. The ionization type is most popular.

An ionization alarm has a small, internal chamber containing a very small amount of radioactive material, which ionizes the air in the small chamber and makes it able to carry an electrical current between two electrically charged electrodes. If smoke particles (excess carbon particles in the air) enter the chamber, they will increase the resistance of the ionized air. This naturally decreases the current flow between the two electrodes, and the alarm will sound when the resistance has increased to, and the current flow drops below, a preset point.

The photodiode-type smoke sensor relies on a different method. In this sensor, a beam of light is projected across a sensing chamber onto a photoelectric cell. When smoke particles enter this sensing area, the light beam is disturbed and the level of light reaching the photoelectric cell is reduced. The alarm is triggered when the light reaching the cell drops below a certain level.

Ionization alarms are most popular because they respond slightly faster to a rapidly spreading fire, which always produces many smoke particles. Both types are reliable, however. They will respond even to the tiny smoke particles produced by a fire before any actual smoke can be seen.

Sometimes a high level of dust in the air can produce a false alarm, and dust accumulating inside the sensor will set the alarm off. Dust in the sensing chamber might also reduce the sensitivity of the alarm. False alarms might also be caused by small insects or high humidity. So, at least avoid

FIGURE 47
Thermal detector

placing a smoke detector too close to a bathroom shower.

Self-contained smoke sensors have integral batteries, usually carbon-zinc or alkaline. The alkaline battery is the most reliable because it lasts longer. All the same, they should be checked at least once a week for safety reasons.

Although a reset button is included in the alarm, any continued presence of smoke will simply result in sounding the alarm again. Therefore, the alarm simply cannot be turned off as long as the conditions that triggered the alarm remain and the device retains electric power.

Smoke detectors can generally be used in any room except the kitchen. There thermal, or heat-sensitive, detectors are used instead (Figure 47). These use built-in pyroelectric sensors able to sense when the temperature in the room rises above a dangerous level. The exact temperature level deemed critical depends on the construction of the sensor. Typically, though, a 50°C (Celsius) level is the most common in Europe, while 135°F (Fahrenheit) sensors are used in the United States, except in kitchens and heating furnace areas, where 80°C (or 190°F) sensors are used instead. The reason is that a higher temperature is normal in these rooms.

This sensor contains a material that melts at a certain

temperature, the chosen threshold temperature, producing electric contact and triggering the alarm.

Thermal sensors can also be wired in parallel, except in kitchens. Their main use is in areas that might produce fires that produce more heat than smoke, for example, with certain types of electrical and chemical fires.

Yet another type of detector is the differential detector. This is an advanced thermal detector for use in environments unsuitable for ordinary detectors, such as certain kitchens and garages. This detector works by adapting itself to the temperature in the area, and it will trigger the alarm when the temperature is rising more than 5 degrees per minute.

For a professional intruder, the main reason to use an existing fire alarm system that forms part of an intruder alarm system is that the fire alarm will override the actual intruder alarm. The intruder will find this very useful because the deliberate activation of the fire alarm might provide an easy diversion. This is another reason why you should not combine the two systems. It is not really necessary to start a fire to activate these systems. Simply blowing smoke into a smoke detector will set the alarm off, afterward, there will be no explanation why the fire alarm sounded at that particular time. The cautious intruder might even take the trouble to blow some dust into the detector to make the alarm appear to be malfunctioning.

The location of the fire alarm will be fairly obvious. Because smoke is impeded to some extent by doors, most fire departments recommend installing several sensors. Optimally, there should be one in every room; at a minimum, there should be at least one in each hallway and one in the kitchen or other area where a fire is likely to break out (e.g., near heating units, fireplaces, fuse boxes). The detectors are almost invariably mounted on the ceiling, generally at the center of the area they protect.

Do not forget to test your smoke detectors and fire extinguishers frequently—if possible, once a week—and replace the battery in each detector at least once a year. While doing so, you should also use the opportunity to clean the detector to ensure a longer working life and make the unit less prone to false alarms. Far too many people invest in fire alarms only to promptly forget about them.

A closed door may temporarily resist a fire. If you wake up because of a fire alarm, do not open your bedroom door until you have the door's temperature. If the door or doorknob is hot, do not use that exit because the corridor outside will be on fire. Use an alternative exit to escape, crawling on the floor if necessary and if there is time. Smoke and heat from the fire rise, so there will be less smoke and not such a severe temperature at floor level. A burning house can be very dark, so keep a flashlight at hand. *Never use an elevator to escape.*

CHAPTER 6

The Most Reliable Locks—and How Intruders Pick Them

The purpose of a lock has been obvious since the days of ancient Egypt, when the mechanical lock and keys as we know them today were invented (Figure 48). You want to lock a door to keep other people from opening the door, whether to disturb you or steal your belongings. This sounds so obvious that it should not need to be stated, but you should remember this old maxim. This is exactly what a lock is for—and nothing more.

The best lock on the market cannot keep an intruder from smashing your door, or your wall, into pieces instead of opening the door. Conversely, the cheapest lock on the market is able to keep other people from entering, as long as their business is legitimate. And in the unlikely case that you are dealing with somebody who is too sophisticated to rely on brute force, the best lock can be picked just as well as the

FIGURE 48
Ancient Egyptian iron keys

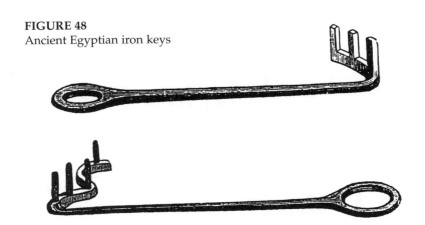

cheapest, if enough time is available. This chapter will teach you what types of lock are safer than others, as well as how intruders are still able to pick them. By using this knowledge, you will have a better chance to select a lock that is neither inadequate nor expensive for your needs.

Lock picking can be defined as the method of opening a lock mechanism by the intrusion of special tools other than the regular key. Lock picking is not easy, and some locks are definitely more difficult to open than others. However, it is possible to open *any* lock without its own key. Although the average burglar knows little of lock picking, the skills are not unknown among locksmiths and more sophisticated burglars.

The reason lock picking is possible at all is that there are always certain tolerances built into the design of the lock. The various parts of the mechanism never fit perfectly. There is always some diminutive empty space in which the lock-picking tools can be inserted.

An expensive lock is usually, but by no means always, designed and manufactured with less tolerance than a cheaper lock. Less tolerance mean less space to insert lock-picking tools. A cheap lock is almost always easier to pick because there are tolerances large enough to insert whatever tools are required to pick the lock. No lock is absolutely pickproof, although there are numerous types of locks that are extremely difficult to pick. In some cases, the level of difficulty is so high that the lock is effectively impossible to pick under field conditions.

The price of the lock dictates the number that will sell, so the vast majority of locks are fairly cheap and consequently easier to pick. For this reason, there is a very good chance that the lock already installed in your door is one of the cheaper kinds.

Devices used as locks today can be divided into the following general types:

- Warded locks
- Lever tumbler locks
- Disk tumbler locks
- Pin tumbler locks
- Tubular cylinder locks
- Magnetic locks
- Combination locks

Locks of the types mentioned above appear in the following shapes:

- Luggage locks
- Padlocks
- Vehicle locks
- Mortise locks
- Surface-mounted auxiliary locks

The last two categories are the two general shapes of door locks and are found all over the world.

Although these general types contain numerous design variations, all locks fall into one of these groups unless the locking device represents a melding of two different types of locks.

Luggage locks, padlocks, and vehicle locks will be described more fully later. Door locks need to be described in more detail because, despite everything, they make up one of your main lines of physical protection.

There are basically two types of external door locks, rim locks and mortise locks (Figure 49). Rim locks are surface mounted (i.e., screwed to the surface of a door). They usually have a spring-operated beveled latch bolt that automatically springs back when the door is shut to hold the door closed. The door is opened by turning back the latch, by using a key or an internal knob, or by moving a sliding handle. In most rim locks, the latch bolt is checked by a safety catch so that the door can be shut without latching. Rim locks are also known as locking bodies.

Mortise locks are mounted inside the door, fitting into a mortise or slot in the leading edge of the door. They are very neat and slightly stronger than rim locks. However, it is generally the wood rather than the lock that gives way during a break-in, so this means little. A rim lock is more commonly fitted to a thin door, since the door is not weakened as much. For the same reason, the staple of a rim lock will not weaken the door frame since it is surface mounted. However, the staple itself is held only by screws and can be broken away from the frame.

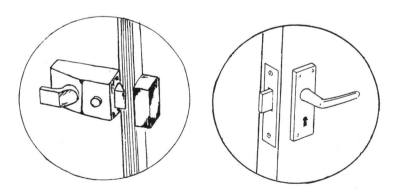

FIGURE 49
Rim lock (left) and mortise lock (right)

The simplest rim lock is the night latch. This is an auxiliary lock with a spring latch bolt that holds the door closed. The night latch functions independently of the regular lock on the door and cannot be dead-locked. The latch is operated by a key from the outside and the knob from the inside. In this lock, the spring bolt can be pushed back with a piece of flexible plastic, as was described in an earlier chapter. In practice, however, most simple rim locks today have a dead-locking mechanism operated by a small thumb piece from the inside that prevents the spring bolt from being forced back. (A common burglar's method is to break a small adjacent window to reach through and open up the lock from the inside.)

There are other types of surface-mounted auxiliary locks, including dead latches (locks that can be automatically or manually locked against end pressure when projected) and surface-mounted cylinder locks used separately from another lock unit. Door chains, surface bolts, and chain bolts are also usually counted as surface-mounted auxiliary locks.

More advanced rim locks, such as the ones with an automatically dead-locking latch (automatic dead lock) or a manually dead-lockable latch, always have a spring bolt with a mechanism that prevents the bolt from being forced back when the door is shut. The internal handle can then also be dead-locked either from the inside or the outside with a key. The lock cannot then be opened except with a key or by picking it.

Rim locks are usually operated by pin tumbler cylinders. A mortise lock is either operated by a pin tumbler cylinder (Figure 50) or flat levers in the lock case itself (Figure 51). The levers are of different heights to correspond to the cuts in the key. The number of lever tumblers determines the security level of the lock: the more levers there are, the greater the number of key variations and the more secure the lock. Five-lever locks are common, but in many countries the nine-lever lock is considered the standard.

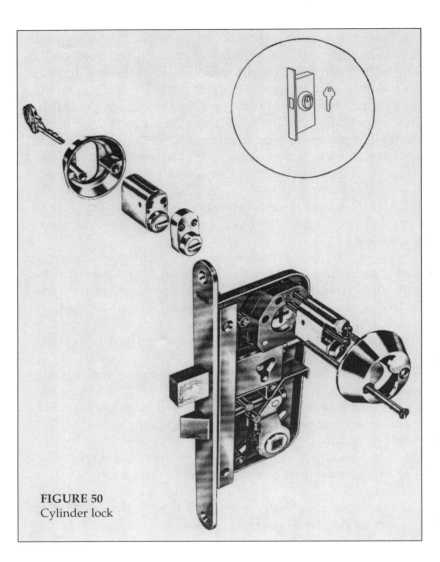

FIGURE 50
Cylinder lock

Some mortise locks also rely on cylinders—for instance, if the owner wants to have the same key for both the front and the back doors. Of course, lever-type mortise locks can also be designed to have identical keys.

No part of the body of a mortise lock is visible when the

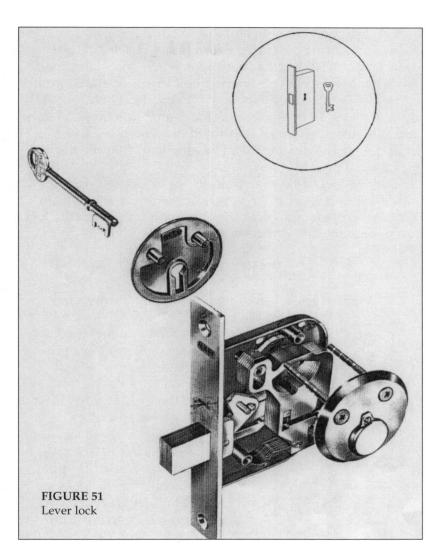

FIGURE 51
Lever lock

door is closed because it is concealed inside the thickness of the door. For this reason, the mortise lock is generally not fitted to doors less than 44 millimeters thick.

There are basically two types of mortise locks: the key-operated dead bolt and the two-bolt mortise lock, or sash bolt. A dead bolt is a lock bolt that has no spring action; it is always actuated by a key or a turn knob. This lock is a true dead lock. A sash bolt has both a latch bolt and a dead bolt. The latch bolt is a beveled spring bolt that is operated from either side by the door handle, while the dead bolt is operated by the key. Two-bolt mortise locks are often fitted on back and side doors, while key-operated dead bolts are fitted to front doors. The dead bolt can be operated from the inside by a thumb piece.

A dead-locked mortise lock cannot be opened from the inside without a key if it has been locked from the outside. The reason is that the bolt cannot be withdrawn into the lock case unless the key is used or the lock is picked.

WARDED LOCKS

Warded locks were first invented by the ancient Romans (Figure 52). The warded lock relies on one or more wards to protect the internal lock mechanism. A ward is a protruding ridge in a lock or on a key designed to permit only the correct key to be inserted in the lock. Warded locks are fairly simple in design and can be found all over the world. These locks are still used in door locks in the older neighborhoods of cities, despite the fact that these locks are not at all secure. Wards are also common in old padlocks. Student locksmiths frequently use these locks to practice lock picking.

Warded door locks are of either the rim or mortise type. Both types operate on the same principle. The sur-

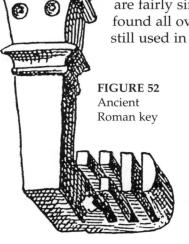

FIGURE 52
Ancient Roman key

FIGURE 53
Type 1—Side ward

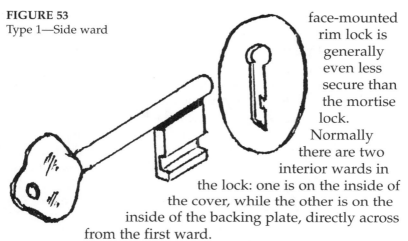

face-mounted rim lock is generally even less secure than the mortise lock. Normally there are two interior wards in the lock: one is on the inside of the cover, while the other is on the inside of the backing plate, directly across from the first ward.

The key for a warded lock is cut to correspond to the number of wards designed into the lock. The key comes in contact with the actual locking mechanism only after it passes all the wards. Then the cuts on the key lift the lever to the correct height and throw the dead bolt into the locked or unlocked position. As long as the dead bolt is retracted, turning the doorknob will activate the spindle and release the door.

The wards are of three possible types:

- Type 1—A side ward is designed to allow only a key with a slot milled on the edge of the key to pass the side ward (Figure 53).
- Type 2—Another type of side ward is designed to allow only a key with the slot milled on the side of the key (Figure 54).
- Type 3—An end ward only allows a key with a slot milled on the end of the key to pass the ward (Figure 54). End wards are commonly milled on both ends, so the key then can be used from both sides of the lock.

The side wards can generally be picked by inserting a

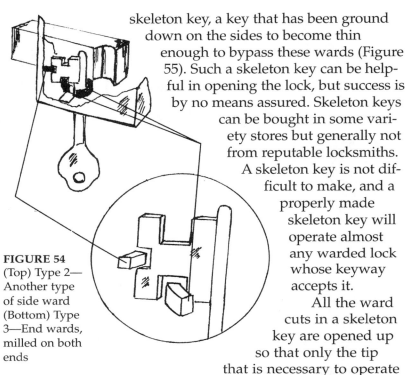

skeleton key, a key that has been ground down on the sides to become thin enough to bypass these wards (Figure 55). Such a skeleton key can be helpful in opening the lock, but success is by no means assured. Skeleton keys can be bought in some variety stores but generally not from reputable locksmiths. A skeleton key is not difficult to make, and a properly made skeleton key will operate almost any warded lock whose keyway accepts it.

All the ward cuts in a skeleton key are opened up so that only the tip that is necessary to operate the latch spring remains. (Most warded locks can also be picked with a T-shaped lock pick, of course.)

FIGURE 54 (Top) Type 2—Another type of side ward (Bottom) Type 3—End wards, milled on both ends

Today, many lock manufacturers try to raise the security level of their warded locks by adding another spring latch with a ward between the two latches. For such locks, a double-headed skeleton key can be used. The principle remains the same, but this skeleton key is designed to take care of that extra complication.

A good locksmith, with plenty of time, can make a duplicate key to the lock by the technique known as impressioning. This method determines the shape of the key by simply studying the lock from the outside. The locksmith inserts a key blank (a key not yet cut or shaped to operate a specific lock) smoked by a candle into the lock. The smoked key blank, when extruded, will show several small marks where the soot has been removed. These marks will tell the smith what cuts to make, where to make them, and how deep they must be. This is a fairly lengthy process and requires some skill.

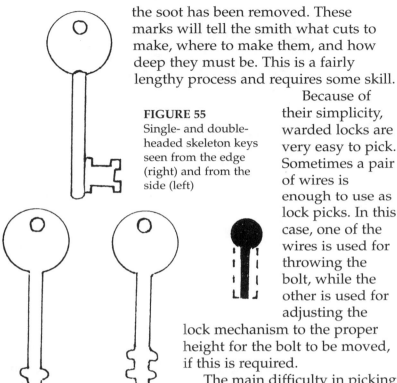

FIGURE 55
Single- and double-headed skeleton keys seen from the edge (right) and from the side (left)

Because of their simplicity, warded locks are very easy to pick. Sometimes a pair of wires is enough to use as lock picks. In this case, one of the wires is used for throwing the bolt, while the other is used for adjusting the lock mechanism to the proper height for the bolt to be moved, if this is required.

The main difficulty in picking a warded lock is not to negotiate the few wards that are obstructing the pick, but to find a set of lock picks of the correct size. As previously mentioned, skeleton keys are often easier to use. Precut blank keys are often used for this purpose instead of regular lock picks.

LEVER TUMBLER LOCKS

The lever tumbler lock, or lever lock, was first introduced in the 18th century. Then, in 1818 in Britain, Jeremiah Chubb invented the Chubb lock, the direct ancestor of the modern lever tumbler lock (Figure 56). Today these locks are still common in light security roles. They are commonly found on desks, lockers, post office boxes, bank deposit boxes, and

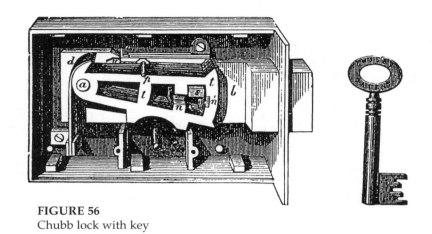

FIGURE 56
Chubb lock with key

similar objects. However, a modified, much more pick-resistant variety of the lever tumbler lock is now also in worldwide use as a high-security mortise lock. In this case, the lock might use as many as nine levers or more.

It is very important to realize that although the security level of the minor lever tumbler locks is lower than for the pin tumbler locks (described below), the security level of the mortised lever tumbler locks is generally significantly higher. These locks are difficult to pick.

A lever lock consists of six basic parts: the cover boss, the cover, trunnion, lever tumblers (usually two, three, five, but sometimes six, twelve, or even fourteen in deposit box locks), bolt, and base. A lever tumbler lock is illustrated in Figure 57. The lock is operated by a standard flat key. After the key has been inserted into the lock, the key is turned, and the key cuts then raise the level tumblers to the correct height. The gates of the lever tumblers will align and release the bolt as the levers are raised to the correct position. The bolt stop is allowed to pass through the gates from the rear to the front or vice versa. This unlocks or locks the lock.

The lever tumbler gates must be perfectly aligned or the

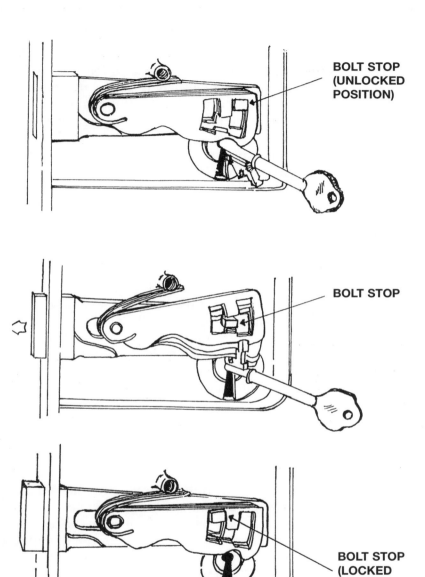

FIGURE 57
Lever tumbler lock

THE MOST RELIABLE LOCKS

lock will not function. Because the key must be cut perfectly, this enhances security.

The key to a lever lock is almost invariably flat. Here, too, an experienced locksmith can make a new key by the impressioning process. The process is much more difficult, however, than is the case of the warded lock. First, the locksmith must do what is generally known as "reading the lock." With a reading tool—which is simply a slightly bent (so that he can see the tumblers) stiff length of wire about 7 or 8 centimeters long, with a wooden handle to make it easier to hold—he can probe the narrow lock keyway. This gives him some idea of how to cut and shape the key.

The locksmith then uses the positions of the lever saddles, those parts of the lever tumbler that are in direct contact with the key, as one clue to the design of the lock. The wider the saddle, the deeper the cut on the key. This process takes considerable skill and long practice. There are also locks that apparently have the same saddle width on each lever, and these locks are even more difficult to read. Here the locksmith must determine the design by finding how high he can raise the various levers.

To pick a lever tumbler lock, the locksmith begins by first inserting the torque (or tension) wrench, a special device used to apply pressure on a lock while its tumblers are manipulated with the pick. He pushes the wrench to the lowest point within the keyway because this will give the pick maximum work space. Then he locates the key notch in the underside of the bolt and applies pressure (Figure 58).

The bolt stop, which is affixed to the bolt, will now bring pressure on the tumblers. By exerting pressure on the lever tumblers with the torque wrench, the locksmith can manipulate the lever tumblers with the pick after inserting it into the keyway. The levers must be moved into position for the bolt stop to move through the lever tumblers' gates.

One tumbler tends to take up most of the tension, so this is the one to work on first. When this tumbler is raised to the right position for the bolt stop to pass through the gate (but

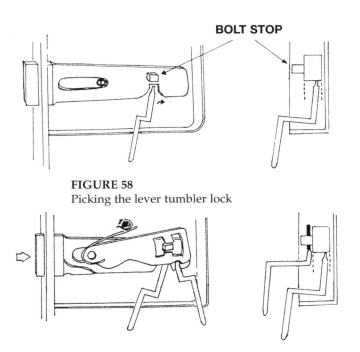

FIGURE 58
Picking the lever tumbler lock

not too high or this will be impossible), the locksmith will feel through his wrench the tension from the bolt slacken, as the bolt now attempts to force its way into the gates. This is the point to stop. Then he must repeat the process with the lever tumbler next to the raised one. When all levers have reached this point, the bolt can be made to pass through the gates by shifting the torque wrench against the bolt. This will open the lock.

If the pressure from the torque wrench is relaxed at any time during the process, all levers in the raised position will drop back to their original positions. Therefore, some pressure must be maintained at all times. Care must be taken not to raise the lever tumblers too high, or they will be above the unlocked position. The locksmith must allow the tumbler to retain its drag as it is raised, because this will help him feel through his torque wrench when he has reached the right position.

THE MOST RELIABLE LOCKS

Simpler desk-type lever locks have two parts that must be moved to open them. The lever, of course, must be raised, but the bolt must also be operated to open the lock. This can be accomplished most easily with an L-shaped lever pick. A locksmith will push back the levers and catch the bolt by turning the pick until he finds it. He will sometimes find it helpful to peer into the lock with the help of a flashlight.

DISK TUMBLER LOCKS

Lever and disk tumbler locks are related in design, although they were invented at different times. A disk tumbler lock, also sometimes known as a wafer tumbler lock, gets its name from its wafer-shaped tumblers, or disks. These locks are now commonly used in garage and trailer doors, as well as in many types of cabinets, desks, padlocks, older vending machines, and cars. These locks can be recognized by the fact that the first flat disk tumbler can be seen through the keyway.

The disk tumbler lock is generally as secure as the lever lock but less so than a pin tumbler lock. It is similar in appearance, and in the broad principle of operation, to the pin tumbler lock even though the internal design is quite different.

The disk tumblers are flat circular or oval-shaped steel stampings that are arranged side by side in slots in a cylinder core, or plug, inside the lock (Figure 59). Every disk has a rectangular cutout in its center that matches a notch on the key bit. The disk also has one or more side projections.

This type of lock employs a rotating core, and this is what makes the disk tumbler lock look similar to the pin tumbler lock. The core is cast so that the tumblers protrude through the core and into slots on the inner diameter of the cylinder. As long as the tumblers are in place, the core is locked to the cylinder. The key, when inserted into the lock, raises the tumblers high enough to clear the lower cylinder slot. They must not be raised so high as to enter the upper cylinder slot, because this will once again lock the plug in

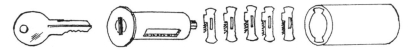

FIGURE 59
Disk tumbler lock

position. When the tumblers have been raised to the right position, the plug is free to rotate. This will operate the bolt.

The key to a disk tumbler lock looks like a cylinder pin tumbler key, but it is usually smaller. Furthermore, it always has five cuts, while a cylinder pin tumbler key might have six or seven.

This makes the disk tumbler lock not very secure. Since every lock has no more than five tumblers, and each tumbler cutout has five possible positions, the design technically allows 3,125 different key changes. In practice, however, some variations are inappropriate, so this leaves us with only around 500 different key changes. Some disk tumbler locks for use in offices (e.g., in desks) are even simpler, with only about 200 key variations possible.

Here, too, an experienced locksmith can make a key by impressioning. First, however, he must read the lock, and in the case of disk tumbler locks this is fairly easy.

First, the locksmith makes a reading tool from a stiff wire. He inserts the reading tool into the lock, so that he can observe the disks in the lock. He raises and lowers each disk by moving the tool until he can see the general positions of all of them. This is generally not difficult, because there are only five variations of the disk tumblers (Figure 60), and the position of them will give him a general idea of the profile to be used for the key. When this is determined, the key can be impressioned in the usual way by inserting a blackened key blank.

Disk tumbler locks, often used in offices, are usually of

THE MOST RELIABLE LOCKS

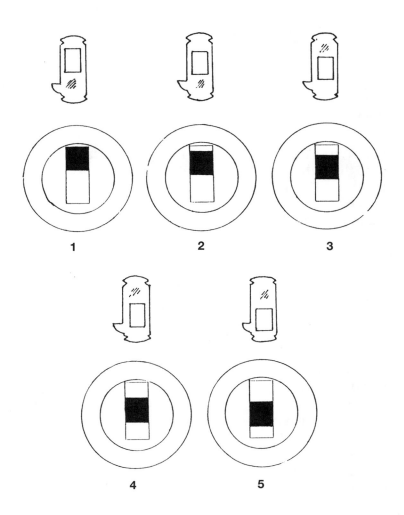

FIGURE 60
The five disk tumbler variations and their positions in the keyway

fairly simple construction. Sometimes a simple sliding-bolt lock is used with the disk tumbler cylinder. In these locks, the bolt is grooved to accept a projection on the back of the

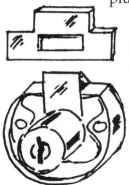

FIGURE 61
Simple disk tumbler lock

plug. The projection engages the groove and converts the rotary motion of the plug into reciprocating motion, opening the lock. In some of these locks (the stronger ones), the bolt-actuating pin is cast as part of the plug (Figure 61). In this case, the plug can usually be released with a probe wire. Locks of these types are often found in drawers and cabinets.

A standard disk tumbler lock (i.e., one that uses a single-sided key) can be opened by the same picks that are required for a pin tumbler lock (these methods are discussed below). The disk tumbler lock, too, can be picked by bouncing the tumbler to the shear line, the space between the cylinder and the plug of the lock cylinder. Usually a rake pick is used for this purpose.

The bounce method is definitely best for picking a double-sided disk tumbler lock, which has disk tumblers protruding through the core in both sides of the cylinder (Figure 62). Such locks are fairly common and sometimes require special lock picks. The important thing, however, is not the pick. The procedure is the usual one, although it has to be done on both sides. After the top disk tumblers have been located and moved to the unlocked position, the locksmith repeats the process with the bottom ones. He will insert the

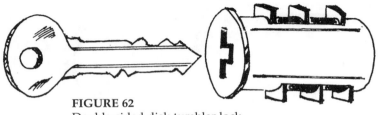

FIGURE 62
Double-sided disk tumbler lock

torque wrench into the keyway and apply a slight pressure to the core as the pick is pulled out.

PIN TUMBLER CYLINDER LOCKS

The pin tumbler lock was first invented in ancient Egypt. The same principle was used centuries later in the well-known Yale lock, introduced by the American Linus Yale more than a century ago (Figure 63). Today the pin tumbler lock is one of the most common types of locks in the world, used for both residential and office building locks, as well as numerous other applications.

A pin tumbler cylinder lock is so named because it relies on pin tumblers. These tumblers are small sliding pins in the cylinder that work against coil springs and prevent the cylinder plug from rotating until the correct key is inserted in the keyway. When the lock is fully assembled, only the plug (the face of its rotating cylinder) can be seen. Locks of this type are generally more secure than the previously described locking devices. They can be recognized by the first pin that can be

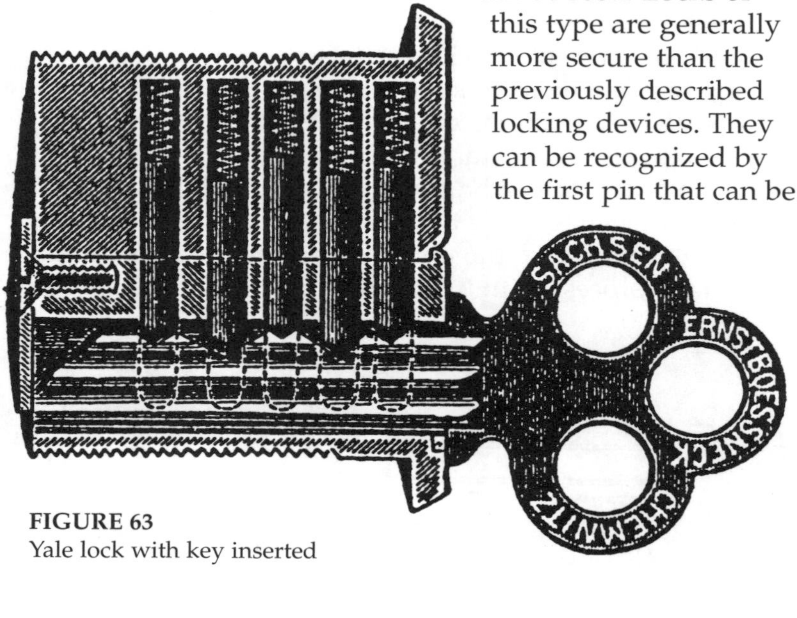

FIGURE 63
Yale lock with key inserted

seen through the keyway. Even the shear point of the pin can sometimes be seen when looking into the keyway.

The pin tumbler cylinder is a completely self-contained mechanism that can be used with a very large number of lock sets. The basic parts of the pin tumbler cylinder are the cylinder case or shell, the plug or core (the cylindrical mechanism housing the keyway), the keyway, the upper pin chambers, the lower pin chambers, the springs, the drivers or top pins, and the bottom pins (Figure 64). All parts of the cylinder are housed by the cylinder case.

The plug is the part that rotates when the proper key is inserted into the keyway. The number of drilled holes across the length of the plug can vary, but there are usually five or six. Some plugs have as few as four or as many as seven holes. These holes are called the lower pin chambers because they each hold a bottom pin. The upper pin chambers are the corresponding drilled holes in the cylinder case directly above the holes in the plug. They each hold a spring and a driver.

The springs and the drivers are usually the same length. The bottom pins, however, are of a different length because they are designed to match the depth of the cuts in the key

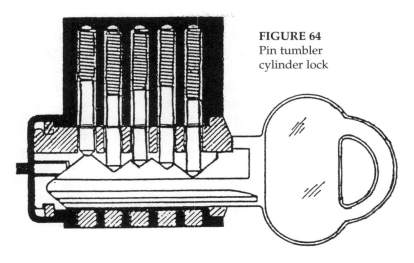

FIGURE 64
Pin tumbler cylinder lock

THE MOST RELIABLE LOCKS

to the lock, by being raised to the shear line by means of the cuts in the key.

Whenever there is no key in the keyway, the springs will press the drivers partway down into the plug, so that it is prevented from being rotated. The plug already holds the bottom pins, so there is not enough room to allow more than the lower portions of the drivers into the plug.

To allow rotation, there is a small amount of space between the plug and the cylinder case. This is the shear line. When a proper key is inserted, it forces the top of all the bottom pins and the bottom of all the drivers to meet at the shear line, and then the plug is free to rotate to the open position.

The plug is generally machined with a shoulder at its forward surface, which mates with a recess in the cylinder. If this is not the case, however, the lock might be able to be opened by shimming the pins with a strip of spring steel. This would force the pins out of engagement and allow the door to be opened. Contemporary locks do not generally allow this possibility.

The pins and the drivers usually have a broken profile to make the lock more difficult to pick. A driver with a broken profile will generally hang up before it passes the shear line (Figure 65). Consequently, a lock with standard cylindrical drivers is easier to pick. For this reason, mushroom (so named because of the shape) and spool drivers are fairly common in pin tumbler locks. A mushroom driver interferes with picking the lock because it engages with the notched cylinder shell when the locksmith attempts to raise the pin to the shear line. The spool driver works in a similar way.

It is quite possible to impression a key to a pin tumbler cylinder. The methods used are different, however, from those used for the warded, lever tumbler, and disk tumbler locks. The main difference is that the key blank cannot be blackened because all the soot would be wiped off when the key is inserted into a pin tumbler cylinder. Instead, the locksmith must depend on the small marks left on the key blank itself

FIGURE 65
High-security drivers:
mushroom (left) and spool (right)

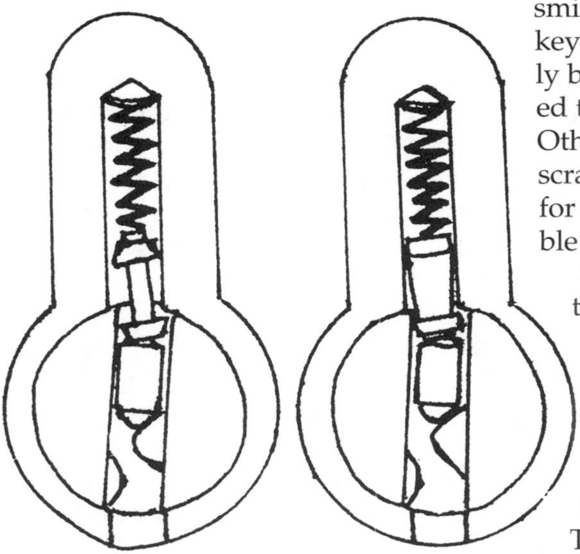

when it is exposed to the pin tumblers. Therefore, the locksmith will polish the key blank thoroughly before it is inserted the first time. Otherwise, the tiny scratches he looks for will be impossible to see.

Picking a pin tumbler lock requires the ability to determine when the cylinder pins have reached the shear line. This can be felt through a tool, or heard as a minute click. A feeler pick can be used to raise the pins to the shear line (Figure 66). The locksmith must be careful not to apply too much pres-

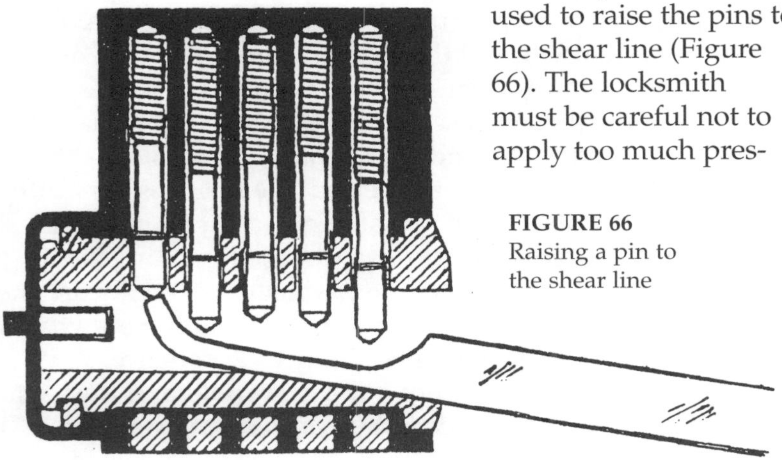

FIGURE 66
Raising a pin to the shear line

THE MOST RELIABLE LOCKS

sure, or the pick will raise the pin above the shear line rather than exactly to it. If this happens, the pin will completely block the attempt to pick the lock open. Furthermore, the pin easily can get stuck at the wrong place.

Before the smith attempts to raise the pins, he will insert a torque wrench into the keyway. The reason for this is that by moving the wrench slightly to the left or right, the drivers are held tight against the plug. Then he will insert the pick and, when all the pins are raised to the proper position, use the wrench with just the right amount of pressure to rotate the plug to the unlocked position and open the lock. He must not to use too much force. Usually only a delicate but firm touch is required to rotate the plug.

It is difficult to first raise and then keep all the pins at the shear line. Fortunately, the fact that the cylinder pin holes in most locks are not in perfect alignment helps the locksmith hold one pin at the shear line with the tension from the wrench while he is working on the next one with the pick.

The first pin to be raised should be the longest one. This is usually also the tumbler that takes up most of the tension. Beginning with the longest pin also allows the locksmith to progress from the smallest amount of pick movement to the greatest. While picking the lock, he'll notice that the plug moves slightly for every pin that reaches the shear line. This movement can also be felt through the torque wrench, making it easier to notice when a pin has been raised successfully.

If one of the pins is raised above the shear line, the smith must release the tension and start again. Less tension is then required when he makes the renewed attempt. It is very difficult to judge the amount of pressure necessary to raise the pins in a lock—even two locks of the same type can react in totally different ways. An experienced locksmith will vary the amount of pressure from light to heavy depending on what is required. Figuring this out takes plenty of experience.

Mushroom drivers present a special problem. It is easy to describe in theory how to pick these locks, but extensive practice is required to do it repeatedly. The secret lies in feeling exactly the moment that the driver is engaging the notched cylinder but before it becomes completely stuck. At this point, the smith releases the pressure on the pin slightly before he attempts to raise it again. He hopes that the driver will have slipped back, so that he can now raise it straight up until it is above the mushroom-shaped trap. When the pin is raised safely to the unlocked position, he increases the tension on the wrench immediately, so that it will not slip down again. Some locksmiths use a spring-loaded wrench for this.

Another way to pick a pin tumbler lock is to use a rake pick or a diamond pick (see below) to bounce the pins to the shear line. This process consists of inserting the pick fully and then quickly withdrawing it while light tension is applied to the plug. This motion often throws the pins apart. The area at the shear line will open up, permitting the plug to rotate. This technique does not work on all types of locks, however.

A rake pick is also sometimes used to rake the lock open, but this process will damage some locks. Another point to consider is that merely forcing the pick rapidly in and out of the cylinder (raking) will only bounce the pins above the shear line. Delicacy is required; if enough skill is not available to pick the lock open, most locksmiths strongly recommend bouncing instead of raking.

A very worn cylinder, especially one with loose plugs, is frequently easy to open with the bounce method. The locksmith usually makes a few attempts at bouncing the lock before he tries to it. After four or five attempts, however, he will no longer waste time bouncing the lock, because it can't be opened this way.

It should be remembered that some types of high-security locks are much more difficult to pick or impression than ordinary locks. The Medeco locks, for instance, are very pick-resistant, because they work by a dual-locking principle. The rotation of the plug in such a lock is blocked by the secondary locking action of a sidebar that protrudes into the cylinder case. The pins have a slot along one side, and the pins must be rotated so that this slot aligns with the legs of the sidebar. The tips of the bottom pins are chisel-pointed, and they are rotated by the action of the tumbler spring seating them on the corresponding angle cuts on the key. The pin tumblers must be elevated to the shear line and rotated to the correct angle simultaneously to allow the sidebar's legs to push into the pins before the plug will turn within the cylinder case.

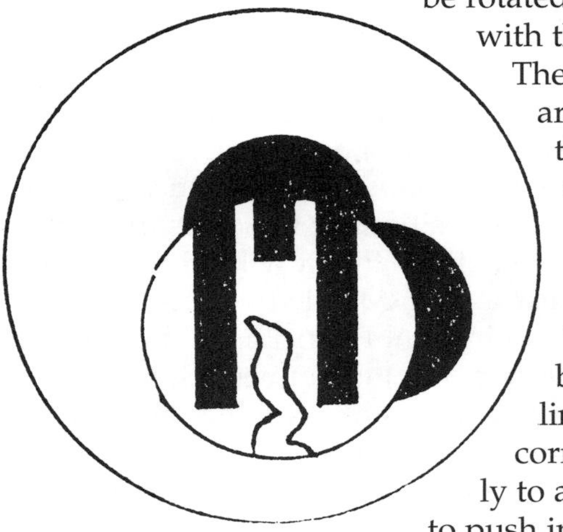

FIGURE 67
Protective inserts on the face of high-security cylinders

Picking such a lock is generally not likely to succeed in the field, even though it might conceivably go well under laboratory conditions.

Medeco cylinders and other cylinders of similar resistance are also protected from physical attacks, including wrenching and drilling, by hardened, drill-resistant steel inserts in the lock. Two hardened crescent-shaped plates within the cylinder case protect the shear line and the sidebar, while hardened rods in the face of the plug and a ball bearing in front of the sidebar protect these areas (Figure 67). These inserts are fairly good for protecting the lock cylinder against drilling, but the door, other parts of the lock set, or even the wall might still be easy to breach by physical attack.

High-security locks also are generally more resistant to impressioning than ordinary locks. Another feature of many high-security locks is that the factory usually maintains control of the key system. The owner must present an ID card and sign a special order form to obtain extra keys. These keys are often known as registered keys and cannot be manufactured without special equipment, usually available only in a price range prohibitive to ordinary locksmiths. The key can only be copied by the lock manufacturer itself or by its authorized affiliates.

TUBULAR CYLINDER LOCKS

The tubular cylinder lock is a variation of the standard pin tumbler cylinder lock. Because the latter locks soon became extremely popular in a short time, the popularity quickly encompassed the tubular cylinder lock as well. Today these locks are generally relied on as high-security locks and, because they are very dependable for their price range, can be found anywhere in a variety of roles, such as key-in-knob locks and desk locks. Yet other common variations are the cylinder locks used on coin boxes, coin-operated washing machines, and vending machines. Larger versions are used as protection on automatic teller machines and in some banks.

One other important application for the tubular cylinder locks is in alarm systems. These locks might be used to protect the control unit or to act as a key switch to the entire system.

The tubular cylinder lock is a real pin tumbler lock that basically works like any ordinary pin tumbler cylinder lock. However, the lock and its key are tubular, and the pin tumblers are arranged in a circle (Figure 68). In this arrangement, all of the pins can be seen from the outside. As is the case with the standard pin tumbler lock, the plug rotates to operate the cam when all the seven or eight pins in the plug have been positioned at the shear line.

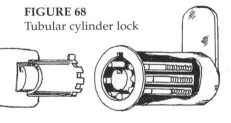

FIGURE 68
Tubular cylinder lock

THE MOST RELIABLE LOCKS

A tubular key has a hollow cylinder-shaped blade that has indentations around the rim of the blade. There are usually seven or eight indentations, corresponding to the number of pins in the lock. Exactly as in an ordinary pin tumbler lock, the pin tumblers are pressed into position by the cut of the key.

It is not easy, but a tubular cylinder lock can be picked with a straight pin and a thin but square-shaped torque wrench (Figure 69). The locksmith will generally have to pick it several times to accomplish the unlocking radius of 120 to 180 degrees. Another problem is that the cylinder will lock after it has been turned slightly. Furthermore, if the lock is left only partly picked, the key will not be able to open it unless the locksmith picks it back to the locked position. This usually takes a considerable time. Another option is to use a special tubular lock-picking tool, such as is used by regular locksmiths (Figure 70).

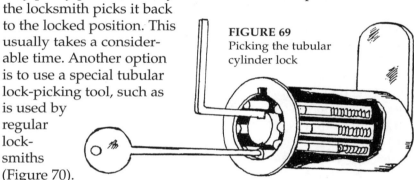

FIGURE 69
Picking the tubular cylinder lock

Many locksmiths will not bother with this, opting instead to simply drill out the lock if the key is lost.

The tubular lock-picking tool is a hybrid tool that both picks and acts as a torque wrench at the same time. Actually, this tool does not pick as much as it impresses the lock. The tool has seven or eight (depending on the type of lock)

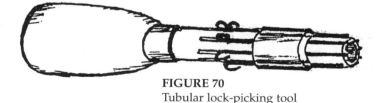

FIGURE 70
Tubular lock-picking tool

steel fingers that adjust themselves to correspond to the cut depth of the original key. These fingers are held in place by a rubber sleeve or a strong rubber band. The rubber band will be tightened or another rubber band will be added once the lock opens so that the steel fingers will remain in the correct position. At this point, the tool can be used either as a key to open the lock or a pattern to cut a permanent regular key.

MAGNETIC LOCKS

Magnetic locks work on the principle that identical magnetic polarities repel each other. In a magnetic lock, there will be a number of small magnets arranged in a certain order. The key contains the same number of magnets, but they are arranged to repel the magnets in the lock. The polarities are arranged in the same way on the key and inside the lock. When the magnets inside the lock are repelled, a spring-loaded bolt will be moved to open the lock (Figure 71).

It is impossible to pick a magnetic lock, but it can be

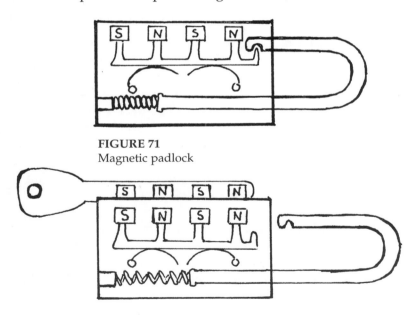

FIGURE 71
Magnetic padlock

THE MOST RELIABLE LOCKS

breached by exposing it to a strong, pulsating electromagnetic field. By repeatedly pulling the shackle (most magnetic locks seem to be padlocks) or the bolt, the lock will then spring open. However, the electromagnetic field is likely to permanently change the magnetic properties of the magnets in the lock. If this happens, the real key cannot open the lock at a later time, so the intrusion will be easily detected.

The electronic field can be created by a portable field instrument if a sufficient power source is available.

It should be noted that magnetic door locks are also used sometimes. They are normally opened by a metal or, more commonly, plastic card containing a magnetic strip that has been coded with a certain magnetic combination. The internal mechanism is the same.

COMBINATION LOCKS

Combination locks used to be associated with safes and bank vaults. Now, however, push-button combination locks and digital keypad locks (Figure 72) are also in widespread use. The digital lock is an electronic lock that can only be opened by keying in the appropriate number code. The push-button combination lock usually consists seemingly of only a control knob and a number of push buttons to control the lock. It is sometimes electronic, but mechanical versions are also common. Nevertheless, these locks are classed as combination locks because they have no key. Push-button combination locks are now becoming more common in hotels, motels, government institutions, and private companies.

In companies and institutions, some high-risk areas can be protected by two devices, one ordinary lock activated after office hours and a keyless push-button lock accessible to the staff during the daytime. Hotels and motels frequently find that combination locks of these types save time and money, so the push-button locks are becoming more popular. Because the combination can be changed every time a guest checks out, security is high.

FIGURE 72
Push-button combination locks (left)
and a digital keypad lock (below)

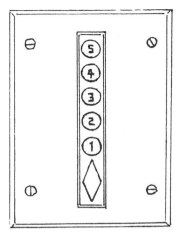

Digital keypads are often used at the common entrance to a block of flats or offices. Unlike some push-button types, these locks are always electronic. The lock is opened when somebody punches the correct code.
Are electronic locks safe? Generally speaking, yes. However, what would be the reaction in a major office building if a van marked with the name of a security or electronic company parks outside during normal office hours, two men dressed in company overalls step out and in full view begin to disassemble ("repair") the electronic lock? Unlike an ordinary lock, electronic locks can be rebuilt to accept *any* number code. Nobody is likely to notice immediately that the lock has been tampered with. Another problem to con-

sider is what happens if there is a power failure. Will you be able to get in and out?

The standard type of push-button combination lock has five push buttons. Pressing the combination in the right order will allow the knob to be moved to open the lock from the outside. This lock can be opened with the combination only.

Another model is the key bypass lock. This lock can be opened by either the combination or the key. Employees or tenants use the combination, while senior management personnel use master keys. Because this lock can be picked and has no special advantages, it might be less common in the future.

Most of these locks, but not all, include an automatic spring latch that locks the door when it is closed. This is to ensure that nobody forgets to lock the door again after he has entered. Another common feature is a faceplate shield to prevent observation of the push-button operation from a distance to learn the combination. The faceplate is the visible part of a lock and therefore the weak link if somebody is observing the area. Yet another option is to eliminate the latch hold-back feature so that the lock can never be kept open. It will remain locked at all times, except for a brief moment when the correct combination is used.

Certain locks are attempts to raise the security level by also allowing two push buttons to be pushed at the same time, in effect producing a different number. However, these locking units cannot use the same push button more than once in a given combination.

The majority of the push-button locks work in the following way. First turn the control knob to the left to activate the push buttons. Then press the push buttons in the correct combination. Finally, release the last push button or push buttons before turning the control knob to the right. This will open the lock. The lock can then be relocked by turning the control knob to the left, or alternatively this may be an automatic feature of the lock.

The lock cannot be opened by removing the control knob,

because this knob is connected to the lock by a friction clutch. The internal mechanism of the lock will be damaged if the control knob is forced or removed.

Other push-button combination locks employ four- or seven-digit combinations. In these locks, there are usually 10 push buttons to choose among. These units can frequently be unlocked by either a four-digit change combination or a six-digit master combination. The master combination might work for several different locks in exactly the same way as a master key.

Combination locks can be electronic, mechanical, or (not as obviously a combination type) remote control. A remote-control lock is activated by an infrared beam from a hand-held device. Remote control locks are generally used only on garage doors and driveway gates. Although the signal opening the lock is supposed to be unique, these locks are not very secure. This makes no difference, since these devices are never used to protect really important positions.

MASTER KEY SYSTEMS

A single key can be cut to match several different lock combinations, thus becoming a master key. Master-keying always relies on coding systems that allow the locksmith to distinguish various key cuts and tumbler arrangements. In most key-coding systems, the tumblers can be set to any of five possible depths. Since most locks have five tumblers and each one of them has five possible depth settings, there can be thousands of different combinations.

Warded locks and lever locks may also be master-keyed, but the security level will be low. Disk tumbler locks and especially pin tumbler locks are more commonly used for this purpose.

When adapted to master-keying, disk tumbler locks are peculiar in that the master key uses a completely different keyway located next to the regular one. This can be seen by closely inspecting the lock. The master key will operate on

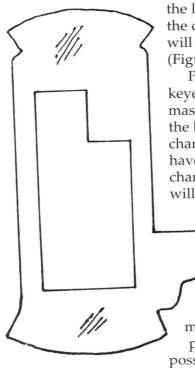

FIGURE 73
Disk tumbler prepared for master keying

the left side of the tumbler, while the change key, the regular key, will operate on the right side (Figure 73).

Pin tumbler locks are master-keyed by adding another pin, the master pin, between the top and the bottom pin in at least one pin chamber. The master key will have some cuts identical to the change keys, but the master key will operate the lock by raising the other pin or pins to the "new" shear line created by the master pin (Figure 74). What actually happens is that there are two breaks for the shear line. In effect, this makes the lock slightly easier to pick, because there are more possible combinations that will raise the pins to the shear line. In master-key systems with many different locks, there might be one or more such master pins between all regular pins. In this case, the system allows for numerous locks and also several submaster keys. These systems are usually factory designed and made, and can include as many as four breaks at the shear line.

A master key will not open all locks in a building unless it is a so-called building master key or an emergency master key, a top-level master key that operates all the locks at all times. It's more common to encounter an engineer's key, a selective master key that is used by various maintenance personnel. Therefore, an intruder must ascertain the level of the master key copied or otherwise obtained before he uses

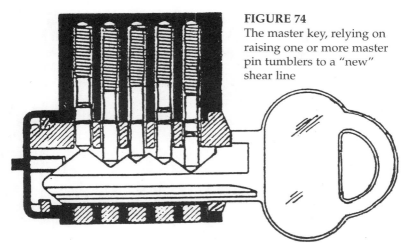

FIGURE 74
The master key, relying on raising one or more master pin tumblers to a "new" shear line

it in an actual break-in. In some low-security systems the level might be stamped on the key, although today it is more common to keep a master key system chart in a safe place, detailing all the various key numbers and the key's position in the key hierarchy. In that case, the key is identified only by a code.

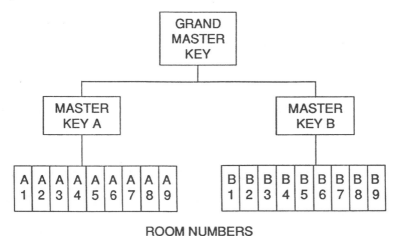

FIGURE 75
An example of a master key hierarchy

THE MOST RELIABLE LOCKS

A master key system is always divided into several different hierarchies, or key levels (Figure 75). An intruder who obtains a master key will attempt to identify its place within the hierarchy, because this will indicate in which locks he can use the key.

COMMON LOCK-PICKING TOOLS

As stated above, few burglars bother to pick locks when brute force works more quickly and easily. Nonetheless, police in several jurisdictions have found it expedient to ban lock-picking tools to all but legitimate locksmiths. Don't be taken in by this illusion of safety. Picking a lock is not easy, and with the exception of the simplest locks, it takes a considerable amount of time. A burglar finds it far easier to rely on an experienced locksmith to make a new key to the lock, especially if he has to enter the area protected by the lock more than once. The locksmith will then use the various impressioning techniques described in the previous chapter. However, lock-picking tools are very easy to make. A few examples should suffice.

A definite prerequisite for lock picking is a very good knowledge of the lock and how its mechanism works. It is also useful to have the proper tools—lock picks—to do the job. A pick is not usually a special tool, but rather any device that can be used to manipulate the tumblers in a cylinder into an unlocked position or to bypass whatever device is protecting the lock from being opened. Lock picks can be easily improvised as long as some basic material and a few simple tools are available. Do not try this, even though it sounds easy: *the possession of lock picks without a proper locksmith's license is illegal in many locations.* On the other hand, you will soon see why a ban on lock picks is not very likely to reduce crime.

A burglar does not really need to carry lock picks on his person except when they are definitely required. Lock picks can be made from easily obtained raw materials almost anywhere, and the only necessary tools are a pair of pliers and a file. These tools, unlike the actual lock picks, do not raise suspicion if someone in possession of them is searched.

In a real emergency, whatever is at hand can be used as a lock pick, although the work will often be slightly more difficult. For instance, it is quite possible to use a large safety pin and a small, slightly bent screwdriver instead of a pin tumbler lock pick and a torque wrench. The burglar simply bends the tip of the safety pin at a 45-degree angle, so that he can use it inside the lock (Figure 76). If no pliers are available, it is easy to bend the safety pin with the help of the keyway of the lock to be picked. The safety pin should be quite large, at least 4 centimeters long, so that it can be easily held while the burglar is working on the lock. It helps if the tip is filed flat, making it easier to locate and raise the pins.

"Real" lock picks do come in numerous shapes, depending on the type of lock to be picked (Figure 77). Lock picks can be easily made from flat, cold-rolled steel, from less than half a millimeter to a little less than a millimeter thick. The actual thickness is really not very important, except that the pick must be able to be inserted into the lock and manipulat-

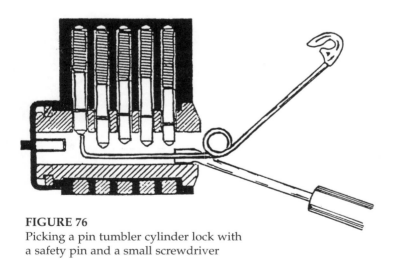

FIGURE 76
Picking a pin tumbler cylinder lock with
a safety pin and a small screwdriver

THE MOST RELIABLE LOCKS

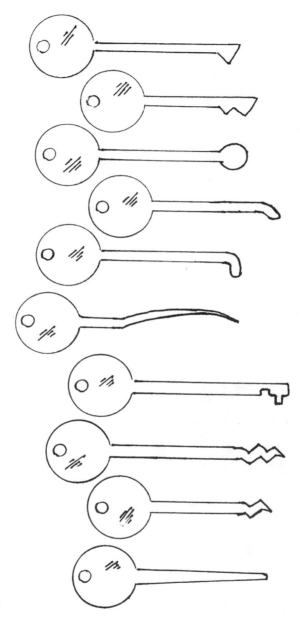

FIGURE 77
Various types
of lock picks

ed in there. The steel strip is typically about 15 centimeters long and at least 8- to 10-millimeters wide. Some prefer to have one end fitted with a handle, while others prefer their lock picks to carry working surfaces on both ends.

A burglar or locksmith makes his lock picks by grinding down the steel strips to the correct size with a file (the best option) or, alternatively, a grinder and a carborundum wheel. If the latter method is used, he takes care that the metal does not become too brittle because of excessive heat.

As mentioned earlier, a locksmith needs lock picks of slightly different design for warded locks (Figure 78). These picks are easy to make. In many instances an old precut key ground down to pass the wards in the keyholes—in effect, turning it into a skeleton key—is sufficient.

One of the most useful types of lock picks is the diamond pick (see Figure 77). It can be used for most pin and disk tumbler locks. Most experienced locksmiths seem to use this pick much more than any other; it is really an all-round pick.

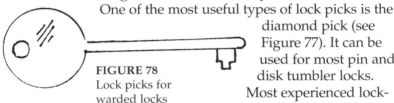

FIGURE 78
Lock picks for warded locks

Usually some kind of torque wrench must be used with the lock pick. These tools are made of hardened spring steel, around 7.5 to 12.5 centimeters in length and a little more than a millimeter thick. Sometimes longer torque wrenches are required because they must be long enough to reach the interior of the lock. There are numerous variations, but the two most useful ones are the basic torque wrench and the steel spring wrench for the really delicate work (Figure 79).

Certain locks (e.g., double-sided disk tumbler locks) require special torque wrenches. Such picks and wrenches are more difficult to improvise (Figure 80) but can be acquired from a locksmith supply house.

The L-shaped lever tumbler lock pick can be made from spring steel no more than 2 to 3 millimeters thick. A torque wrench of the same thickness is also useful (Figure 81).

THE MOST RELIABLE LOCKS

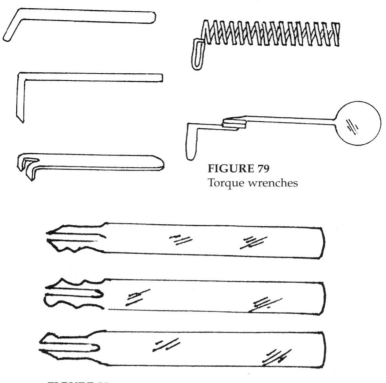

FIGURE 79
Torque wrenches

FIGURE 80
Lock picks for double-sided disk tumbler locks

Some individuals use a rake pick (see Figure 77). As noted earlier, such a pick is not very useful. If a rake pick is really desired, the locksmith can just as effectively use an ordinary diamond pick for the same purpose. Although some people make a living by manufacturing and advertising numerous types of lock picks, it is not really necessary to have them all. In most cases a simple pick will do. A locksmith who cannot pick a lock with a diamond pick is usually no better off using any other type of lock pick.

Finally, the device known as a pick gun (Figure 82) should be mentioned. This hand-held device can be used to bounce the driver out of the plug in a pin tumbler lock and into the cylinder case. No great skill is required to use a pick gun, but this device is not necessarily faster or even better than picking the lock manually. The pick gun cannot pick all types of locks, so it is of more limited value than an experienced locksmith. The only time a pick gun might be really useful is when a locksmith is picking a lock equipped with mushroom or spool drivers.

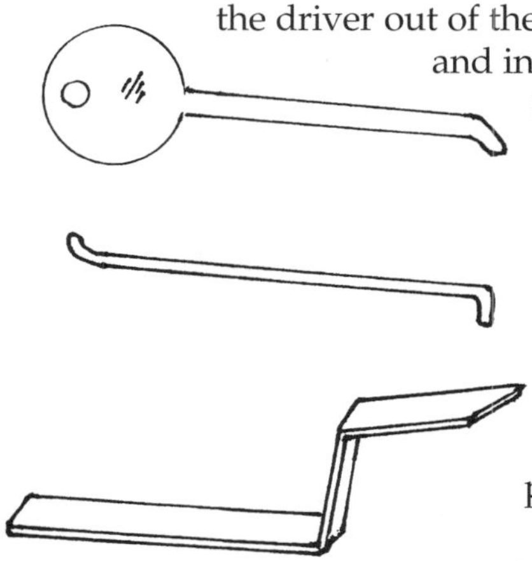

The pick gun is inserted into the keyway so that its pick is barely touching all the bottom pins. When the trigger is pressed, the pick gun raps the pins up to the shear line, so that a turning force allows the plug to rotate and the lock to open.

FIGURE 81
L-shaped lever lock pick, the same pick improvised from wire, and a suitable torque wrench

There is also a manual version of the pick gun, which is known as a snap pick (Figure 83) and can be easily made from spring steel. Just as with the pick gun, the pick portion of the snap pick is inserted into the keyway and held so that it just touches the bottom pins. The locksmith then presses the upper part down with his thumb

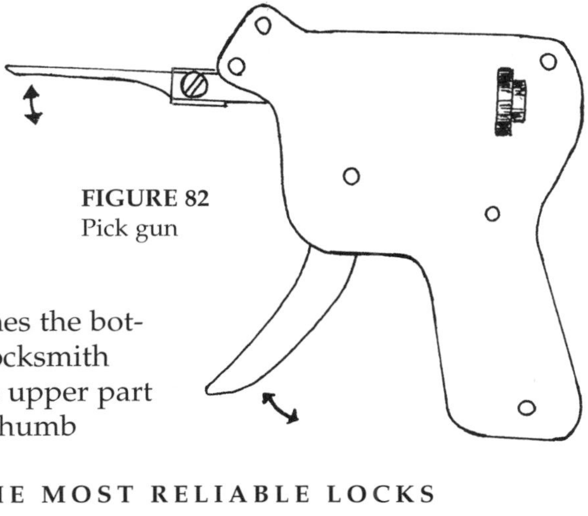

FIGURE 82
Pick gun

THE MOST RELIABLE LOCKS

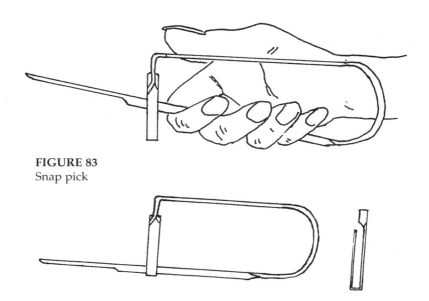

FIGURE 83
Snap pick

before quickly releasing it. This raps all the bottom pins. Because these pins remain relatively stationary, the force will be transferred to the drivers, which are forced to move upward, compressing the spring. Since the area at the shear line will then be open for a split second, a slight turning force will allow the plug to rotate.

The principle of both the pick gun and the snap pick is easily demonstrated by the fact that force is transferred through stationary matter. This can be seen in an experiment with three coins on a table. If the coins are lined up touching each other and the coin in the center is held firmly to the table with a finger, the coin on one side will move if the coin on the opposite side is pushed to strike the stationary center coin. The momentary force is transferred through the stationary coin and moves the opposite coin. In the same way, the drivers are moved by the force applied to the bottom pins.

When working with a pick, an experienced locksmith uses the narrowest pick available to give himself maximum working space. He holds the pick in about the same way as a pencil. No wrist action is desired: only the fingers are dexterous enough to manipulate the pick inside the lock. Wrist action will only make him tired, and lock picking is not a matter of physical strength.

Many locksmiths also find it helpful to steady their hand with the little finger against the door when working on the lock. If the lock is a key-in-knob cylinder, however, they steady the hand against the edge of the knob instead.

The pick should be able to enter the keyway above the torque wrench without moving any of the tumblers. If this is impossible, then the torque wrench is either too high in the lock or the keyway grooves are such that the torque wrench must go in at the very top of the keyway. In these cases, picking the lock is difficult.

Sophisticated burglars have various ways of acquiring lock-picking tools despite the fact that most manufacturers and distributors of locksmithing equipment often quite publically refuse to do business with anyone other than professional locksmiths. Yet, these manufacturers are usually quite satisfied to send their products by mail to anybody who in any way can prove that he is a "professional locksmith." I say "in any way" because I have found out that a fancy letterhead, business card, photocopy of a forged locksmith license, or even a photocopy of an advertisement in the yellow pages is usually enough to convince some firms of a prospective buyer's professional status. After all, money does not smell, and these people are eager to sell.

Many lock pick sets, both commercial and improvised, are designed to be hidden in various objects, so as not to compromise the user. One popular design is to hide the picks inside a pen cover or the handle of a hobby knife. In the latter case, the hobby knife can also be used as a handle for the pick, making it easier to hold. Another place to hide a lock pick set is inside a jackknife handle. Only the burglar's imagination limits the places to hide his lock picks.

Every burglar knows that a real key is quicker and safer

than any lock-picking skills. Even if it turns out to be impossible to acquire one of the keys to the targeted building, there might be ways of at least inspecting the original key. If a key can be handled for some time, even if it cannot be brought out from the building, it is almost always possible to make an impression in wax or a full-scale drawing, which can later be used to make a copy.

The full-scale drawing must be done carefully, with detailed measurements of every part of the key. This is generally easy, though tedious, except when it comes to the diameter of the key. For certain types of locks, the diameter of the key can be measured by wrapping something (e.g., a paper clip) around the original key. The paper clip (or other item used to measure the diameter) can then be taken from the building. In this way no special equipment is required for measuring the key. The lesson you should learn from this is to *never*, under any circumstances, allow strangers to handle your keys. Do not ever leave strangers alone in the room where you keep your front door key, or in the front hall if you keep the key in its lock.

CHAPTER 7

Other Common Locks

In the previous chapter, we examined the most reliable locks and ways they are picked by burglars. In this chapter we will look at other simple types of locks, including suitcase locks, padlocks, and briefcase and padlock combination locks, as well as ways that they can be broken into.

SIMPLE SUITCASE LOCKS

A locked suitcase is no more of a guarantee against intrusion than a locked house. Almost all suitcase locks are of the simple warded type, having only a primitive bolt mechanism to keep the case closed. A few suitcase locks rely instead on a lever-type mechanism. The lever suitcase locks are usually recognized because the key will go half a centimeter or deeper into the lock before turning. A warded lock is much shallower.

It is easy to make a skeleton key that will open most

types of warded suitcase locks. Almost any suitcase key can be used for this purpose.

Alternatively, simple suitcase locks can be picked with a special lock pick (Figure 84), which is easily manufactured from a strip of steel. The lock pick is merely inserted into the keyway, and when the bolt is located, the pick is turned to manipulate the bolt. This will open the suitcase.

Many suitcases have simple combination locks, known as sesame locks. These locks not very safe either and are described below. The bottom line is that you should never put anything valuable into a suitcase. The lock can be easily picked, and if it is checked on an airplane or left unattended for even a moment, anybody can simply carry the suitcase away to be opened at leisure.

PADLOCKS

The padlock was first invented by the ancient Romans. Today a variety of padlocks are in common use, including warded, lever, disk tumbler, pin tumbler, and combination types.

Padlocks can usually be picked easily. A locksmith

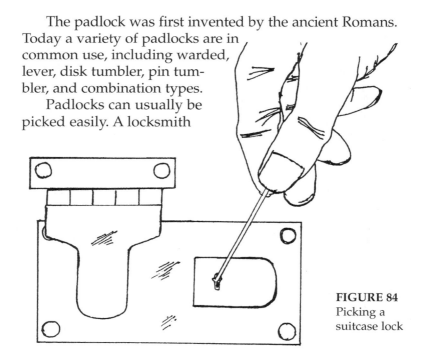

FIGURE 84
Picking a suitcase lock

holds the padlock with the same hand that uses the torque wrench. It is generally easiest to do this by holding the padlock between the thumb and the forefinger. Then he can hold the wrench with his ring and little finger. His main hand is then free to work the lock pick. An apprentice locksmith may need to take some time practicing how to hold the padlock to make certain that it is easy to work on. Of course, different people may prefer to hold the padlock in different ways. All in all, the picking process is not difficult.

Padlocks are not really that different from standard locks of the same type. Some smaller and simpler warded padlocks have a single ward only and take very simple keys. Otherwise, these locks are designed in the same way as other warded locks. The majority of the warded padlocks have three wards, or at least two. The key must pass through the wards before it can disengage the spring bar from the slot in the shackle end.

A warded padlock can be defeated by a skeleton key, which is an ordinary key that has been ground down (Figure 85). There are a few basic shapes, one of which will almost certainly defeat the lock. NOTE: *Skeleton keys are illegal in many locations.*

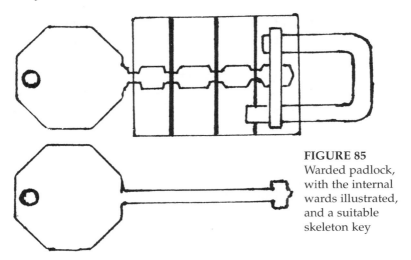

FIGURE 85
Warded padlock, with the internal wards illustrated, and a suitable skeleton key

OTHER COMMON LOCKS

FIGURE 86
T-shaped wire pick

Most warded padlocks can also be picked by an improvised T-shaped wire pick (Figure 86), which can be easily mode from a piece of stiff piano wire.

An old method of picking padlocks is to insert a hatpin through the keyway and use it to disengage the shackle bolt. This still works with older types of padlocks but not with the newer, more secure ones.

Warded padlocks can also be impressioned easily. The procedure is the same as with ordinary warded locks.

The other types of padlocks, including the cylinder types, can also be picked or impressioned in roughly the same way as ordinary locks of their type. Disk tumbler padlocks can also be picked, but it takes practice. Another measure that sometimes works is to acquire a set of test keys from a locksmith supply house: such keys will facilitate getting the lock open.

Whenever picking a padlock, the locksmith must remember that in some padlocks he needs to pull the shackle to help release it from the locking spring. If this does not help, then he must repeatedly work the shackle in and out while picking the lock. This will eventually unlock the mechanism.

Every burglar knows that it is far easier and faster to attack either the padlock or, more commonly, its staple with heavy-duty tools. Again, a padlock is no safer than its surroundings.

BRIEFCASE AND PADLOCK COMBINATION LOCKS

Sesame locks (Figure 87) are extremely simple combination locks frequently found on briefcases and suitcases. These locks consist of three dials numbered from zero to nine. The number of possible combinations is very low, only

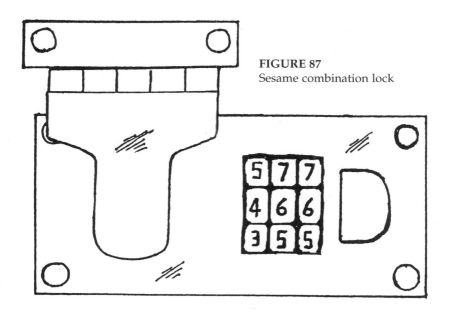

FIGURE 87
Sesame combination lock

1,000. This means that the correct combination can be found by trial and error if there is enough time. An experienced intruder expects to use about 30 minutes to go through all possible combinations, but it usually takes less time. The correct combination is often found long before he has checked all the possibilities.

The intruder begins the process by setting all three dials to zero. Then he moves dial C to 1, 2, 3, 4, 5, 6, 7, 8, and 9, one after the other, while pressing the catch after each change in number. Unless the correct combination has been found, he will change the setting of dial B to 1. Dial A will remain on zero. Once again, he repeats all numbers on dial C, while pressing the catch as before. If the correct combination has not been found, he will set dial B to 2 and repeat the process. He will continue the process until dial B has been set to all possible numbers. Then he will set dial A to 1 and once again repeat the entire operation. Dial A is then set to 2, and the process is repeated.

OTHER COMMON LOCKS

This process, although boring, is easy to perform and sooner or later results in the lock's being opened. Most intruders would not try to think about what they're doing; they have simply learned the process by rote and do it. That is, unless they give up and simply slash the briefcase with a knife.

Combination padlocks with a dial can be opened in basically the same way as other dial-type combination locks (as are used in safes; see Chapter 9), but they are usually simpler in construction and, consequently, easier to open. For this reason, they are often used for practice purposes. It is easier to locate the gates inside the combination padlock if the locksmith pulls out the shackle at the same time that he rotates the wheels. It is also easier to open combination padlocks that have been in use for a long time. The gates on the wheels have become smoothed down, which simplifies the manipulation of the wheels.

Yet another type of combination padlock is the sesame padlock (Figure 88). This padlock has no dial; instead, it works on the same principle as the sesame lock described above. However, the sesame padlock has four combination wheels, numbered from zero to nine. The method for opening a sesame lock with only three wheels is not very practical on this padlock, because the time needed to check all possible combinations would be approximately 10 times as long, or five hours. Of course, there is another way to open this lock.

The four-wheel sesame padlock is designed to unlock the shackle only when each wheel is positioned so that a flat

FIGURE 88
Sesame padlock

spot on each wheel is aligned with the corresponding flat spots on the other wheels. Each wheel has a changeable hub with this flat spot. When aligned together toward the side of the lock stamped with the trademark, these flat spots will unlock the padlock. The current combination of the padlock determines where the flat spots will be located on the wheels.

A special tool (Figure 89) made of very thin steel can be used to locate these flat spots. An intruder inserts the tool into the lock between the wheel and the housing. He turns the wheel slowly and tries to locate the flat spot. When the flat spot is found, he will either add or subtract five from the indicated number, thus getting the correct combination number of that wheel. This is not a particularly safe lock.

FIGURE 89
Manipulating the sesame padlock

OTHER COMMON LOCKS

CHAPTER 8
Reinforced Doors and Security Bars

As stated earlier, given the choice between good doors or windows and good locks, choose the former. A strong door or window secured with bars will beat even the most sophisticated lock as a theft deterrent.

REINFORCED DOORS

The best way to protect an upper story apartment is almost invariably to replace the standard door with a special reinforced steel door and frame. If you cannot do this, you can at least reinforce your regular door by driving steel bars through it or mounting steel plates on the inside. Do not forget to also reinforce the frame with thick, unyielding steel bands inlaid in the frame, as well as the other improvements mentioned in Chapter 1. Even though a determined burglar might still be able to break through such a door, it will be awfully hard to do so. He is likely to give up and go on to a less well-protected neighbor's door.

Always reinforce your door and its frame before you make any additional investments for home-security purposes. Attractive steel security doors are available, so it's not necessary to spoil the outward appearance of your home. In a high-rise building, such a door with an appropriate frame may well prove sufficient protection against all but the most sophisticated burglar.

SECURITY BARS

The existence of security bars poses a very serious problem for the average burglar. Such bars, used as reinforcement next to an existing door, or the bars used as a decorative grille for a window, are very difficult to remove except through brute force or the noisy use of a carborundum wheel. Most burglars will not even attempt to break through them. Grilles, especially large sliding grilles (Figure 90), are commonly used in many countries for protecting large areas of glass or doorways. Other grilles are of the detachable type, locked in place only by fixed locks or padlocks. If this is so, the locks are the weak links in the security system.

Some older buildings rely on solid wooden shutters instead of steel grilles. Shutters, especially wooden louvered ones, are easy to break through, so they present no special protection.

After a reinforced door and door frame, and preferably used in conjunction with them, the best things to install are security bars to protect all windows and doors, at least on the ground floor and wherever else burglars may reach by climbing nearby trees or structures.

Common complaints against security bars are that they make cleaning the window difficult, which is true, and that they are ugly. However, grilles today come in a wide range of designs, most of them very decorative. There is no need to invest in bars that make your home look like Alcatraz. Besides, by putting the grille inside the window, there is no risk that the bars will rust. The window will also be easier to clean or repair, at least from the outside.

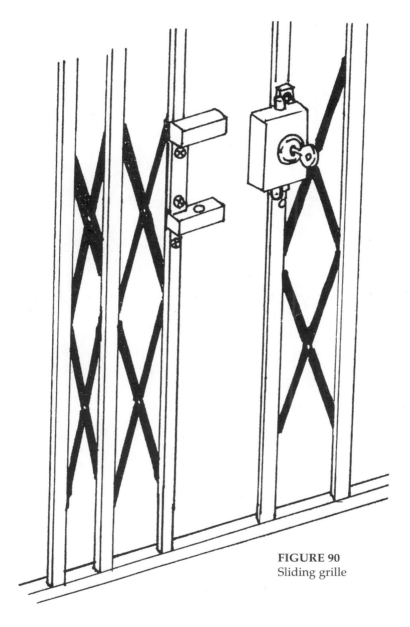
FIGURE 90
Sliding grille

CHAPTER 9

How to Protect Your Safe

There are many different types of safes. Some of them are built to resist burglars; others are designed to resist fire. And some are designed to resist both burglary and fire.

SAFES AND HOW BURGLARS CRACK THEM

Most safes can be cracked without too much trouble, although the contemporary ones are more difficult than the older models. The burglar's main problem in cracking a safe is creating a lot of noise. Furthermore, most safes are too heavy to move from their location, so the work must be done in place on the premises. If the safe is small and light enough, and not securely fastened to the wall or floor, the burglar can remove it and open it at his leisure. Such a safe is nothing but a toy.

Floor Safe

The most reliable type of safe in a house is the small, underfloor type because this type is out of sight and can be permanently set in a reinforced concrete floor (Figure 91). A floor safe is recessed into the floor so that the lid is just below or at the floor level. The floor safe is made of thick steel with a small but strong lid. Different sizes are available, depending on the depth available under the floor, but the opening will almost always be relatively small.

A safe of this kind is preferably placed in a corner near the wall so that it can be covered with the carpet or linoleum and still be reached with a minimum of inconvenience. A corner location also gives the intruder less room to work, even if he happens to discover the safe.

A floor safe should not be installed in a room where the floor is likely to get wet, such as a bathroom or a laundry room. Even though it is usually hidden under the carpet, other hiding places should not be disregarded, for example, under a false floor in the base of

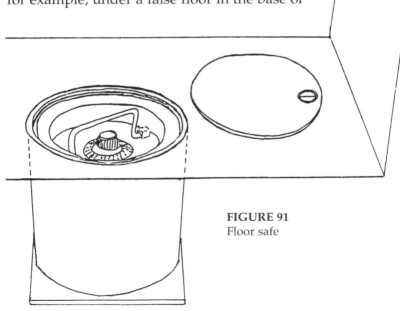

FIGURE 91
Floor safe

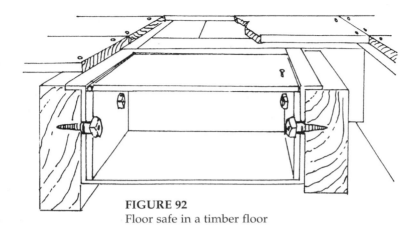
FIGURE 92
Floor safe in a timber floor

a floor cabinet, which can easily be located even in a room with a tiled floor.

Floor safes can be fitted in a suspended timber floor (Figure 92). These safes are in effect secure boxes, often with combination locks, that fit between adjacent joists and bolt or screw to them. The screws or bolts are fastened from the inside of the safe. The floorboards should be removable to reach the safe. The safe itself is generally also covered with a sheet of plywood or hardboard to make the surface level with the surrounding floorboards.

Wall Safe

In a block of flats, the equivalent of the floor safe is the wall safe (Figure 93). This safe is also easily concealed. A wall safe is a small security box set into the wall, replacing one or more existing bricks. The size of the wall safe is measured by the number of bricks it replaces. Wall safes are commonly one, two, or three bricks high and one brick deep, although double-depth models are also available.

Even though wall safes are sometimes hidden very cleverly, one is often amazed to see how often the safe is located behind a painting. This is the first place burglars look for a

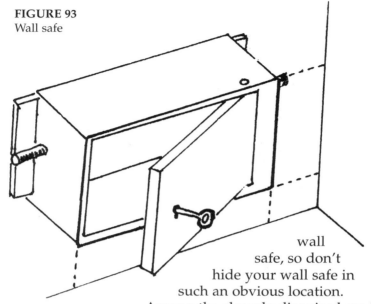

FIGURE 93
Wall safe

wall safe, so don't hide your wall safe in such an obvious location.

Among the cleverly disguised models, there is even a commercial cashbox-sized wall safe made to look like an electrical power socket (Figure 94).

A wall safe is easier to breach than a floor safe because the latter is more difficult to open or lever out of position. It

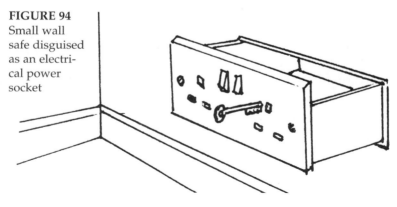

FIGURE 94
Small wall safe disguised as an electrical power socket

is also quite easy to knock out a masonry wall, especially in an older house where the mortar is crumbling. The bricks around the safe can be chopped out with a bolster chisel. The wall safe is then removed and brought to a safe place to be opened.

Strong Room

In certain buildings, it may be preferable to install a strongroom instead of a safe. These are sometimes difficult to break open, but the principles are the same as with ordinary safes. A strongroom is generally less secure than a bank vault, which is very difficult to force open.

Bank Vault

A burglar applies the same methods used for cracking ordinary safes to break into bank vaults. Another interesting point about a bank vault is that it often has a hidden emergency entrance in the shape of a hatch. This is to ensure access even if the main door has become impossible to open because of a failed burglary or a mechanical problem. This hatch is also heavily protected, but because it is smaller and usually hidden behind a steel plate, it is not as well protected as the main door.

Free-Standing Safe

Free-standing safes, especially if not prohibitively heavy, can usually be secured to the floor and/or the wall. Unless the safe is secured in this way, the burglar's best option is to simply remove the safe and crack it open in some safe spot. Every type of safe is most easily cracked with the help of a carborundum wheel. The disadvantages of this method are that a steady supply of electricity is required and it makes a terrible noise. Fortunately, sufficient power is usually available in most houses, although it is prudent to bring an extra set of fuses because the owner may have removed the regular ones to prevent this method from being used on the safe.

Bionic Safe

Some companies now market what they call a bionic safe. This is a safe with an integral alarm system, as well as a few other protection devices. The sensor is usually an inertia sensor that activates a siren and at the same time—through an automatic dialer connected to the nearest telephone—dials a preprogrammed telephone number.

If the safe is cracked despite these precautions, a self-contained explosion will destroy all materials inside. This is designed to cause no harm to any people or property in the vicinity. Furthermore, if the safe is opened without authorization, a sudden burst of high-powered light is emitted to disable the intruder with a temporary stunning and blinding effect. This type of safe is not in widespread use.

• • •

As a final note, it should be remembered that many safes are designed with an outer construction that imitates high-quality wood paneling. This is both to hide the safe and make it fit inconspicuously into a home or office, the best protection always being to hide the safe well. When searching a home or an office, a diligent intruder is careful to check all possible hiding places for a safe, however small they might be. Do not let him find yours.

Common Techniques for Cracking Safes

Many types of safes rely on combination locks. These can be manipulated to the open position, but this is difficult (the technique is described in the next chapter). Combination locks can also be opened by drilling. The lock cannot be completely opened by simply drilling, but the drilled holes will help the intruder manipulate the wheels of the combination lock.

If this method is chosen, then the burglar drills two 3-millimeter holes in the back of the lock. After this, he turns the dial and observes what happens inside the lock through these holes. The gate can be seen through the hole with the help of a flashlight, or a piece of piano wire can be inserted through the hole as a probe. Whenever he has found the gate and aligned it with the hole, the intruder can note the number on the face of the combination dial. Then he determines the distance, expressed in divisions on the dial, between the bolt and the gate aligned with the hole. By subtracting this distance from his reading, the burglar ascertains the combination number for that particular wheel.

After the first number of the combination has been obtained, the burglar reverses the rotation of the dial and repeats the process with the second wheel. The combination of the third and the fourth wheels, if the latter is present, can be determined in the same way.

There are also cruder ways of cracking a safe, such as with explosives. In this case, it is most common to drill a hole above the dial and insert a finger of a glove or a small plastic bag with explosives in the hole. An explosion here will destroy the lock mechanism, but it by no means guarantees that the door will open because of this. It is fairly common for the door to jam as a result of the explosion. Another method, almost as old, is to use nitroglycerin. The burglar first drives a steel wedge into the top seam of the door. He then allows the nitroglycerin to seep around the inner edge of the door, so that the door will be blown off the safe when the nitroglycerin is ignited.

None of the methods requiring explosives are much used today because they are both dangerous and noisy. A neighbor might be disturbed by the sound of a carborundum wheel, but this alone is usually not enough to make him or her call the police; after all, many repairmen use this tool. An explosion is quite a different matter.

Another way of cracking older, inferior types of safes is to drill a hole in the corner of the front plate of the door. This plate can be torn away with the help of a long crowbar to expose the lock mechanism. The lock mechanism can then be easily manipulated open. This method is completely ineffective against a modern safe, however, because the door is massive.

HOW TO PROTECT YOUR SAFE

Most of these older methods (e.g., drilling) won't work on today's safe doors that are laminated with hard steel and beryllium-copper plates. One method that still works, though, is what is generally known as a "torch job." An oxyacetylene torch is used to cut through the safe, but this requires bulky equipment and special training. This method is not really common today because a carborundum wheel does the same job more easily and safely.

Most older types of safes relied on fairly thin metal walls padded by an insulating material. However, over time, the insulating material had a tendency to compress, until the upper parts of the walls were completely empty and, therefore, very easy to break through.

COMBINATION LOCKS

As noted in the last chapter, many safes rely on combination locks. A combination lock is a lock that may or may not be operated with a key but always can be operated by entering a combination of numbers or other symbols. This is done by either rotating a dial or pushing buttons.

The most well-known combination lock is the safe combination lock, which consists of a series of interconnecting wheels that rotate around a central core. The device is controlled in its revolutions by an outside combination dial.

All combination locks of the safe type operate on the same principle, even though there are internal differences between different types of locks. An internal wheel pack is rotated by manually turning the external combination dial. The wheel pack consists of a series of interconnecting wheels, or tumblers, usually three but sometimes four. The dial has 100 different numbers, or positions, and a three-wheel mechanism has almost 1 million different possible combinations. The four-wheel combination locks have almost 100 million possible combinations. Each wheel is designed to align its gate with the bolt-release mechanism only after a certain number of revolutions and a certain degree of rotation. This design can be programmed easily, so that the combination that opens the lock can be changed according to the owner's wishes.

Basically, a combination lock of this type works in the following way. In addition to the wheels, the complete mechanism consists of a dial, spindle, and driving cam (Figure 95). These devices form the driving mechanism that moves the wheels into a locked or unlocked position. The wheels are moved by driving pins that are affixed to the back of the driving cam as well as to the wheels. These driving pins engage and disengage the wheels as the cam revolves according to the movement of the dial. When the dial (by way of the cam) has set all wheels in the right position—and when revolved slowly either back in the direction opposite to the last combination number or (in newer safes) by once again changing

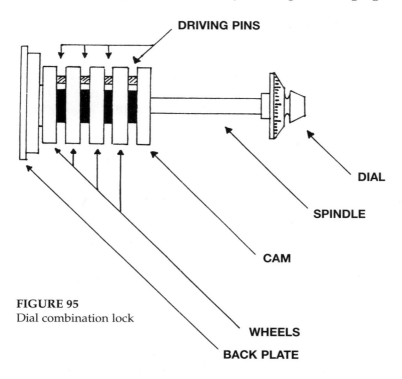

FIGURE 95
Dial combination lock

direction—the dial will operate the bolt of the lock to the open, unlocked position.

The wheels are always rotated in order, and the number of turns depends on the number of wheels. The most common lock type, the one with three wheels, requires a 3-2-1 rotation sequence, while most high-security combination locks have four wheels and require a 4-3-2-1 sequence. The combination dial is first always rotated (in the basic model) three turns, then in the reverse direction for two turns, and finally again in the reverse direction for one turn. Most four-wheel combination dials are designed to begin rotating to the left, but this is not universally true. Likewise, most three-wheel combination locks are designed to begin rotating to the right. Of course, there are differences between different types of safes.

As the wheels are rotated, the gates will be aligned by stops, one for each wheel and one on the wheel-pack mounting plate. The bolt will be free to release only when all gates are aligned.

Older and inferior types of combination locks can be distinguished by audible clicks when the wheels rotate. This allows a skilled individual to manipulate the lock without knowing the combination in advance. Contemporary combination locks have at least three false gates in every wheel (Figure 96), so that manipulation is much more difficult. It can still be done, but only by an expert who has lots of practice, as well as special training, and knows the peculiarities of that particular type of lock he is working on.

Manipulating a combination lock in this way is a matter of touch as well as hearing. The latter is usually assisted by an electronic stethoscope, but the sense of touch can only be developed by long training. The process can be described in the following way, although considerable experience is necessary to succeed.

An intruder rotates the combination dial slowly until he hears a very faint click. He then feels that the bolt is hesitating, touching the *far side* of the gate. At this point, he moves

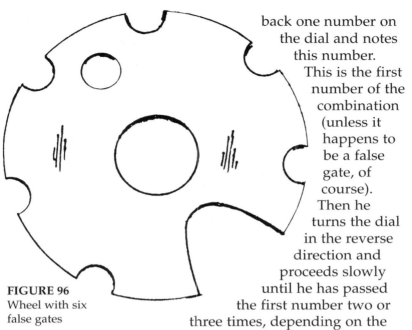

FIGURE 96
Wheel with six false gates

back one number on the dial and notes this number. This is the first number of the combination (unless it happens to be a false gate, of course). Then he turns the dial in the reverse direction and proceeds slowly until he has passed the first number two or three times, depending on the type of lock he is working on. As he slowly continues to turn the dial, he will notice that the bolt touches the far side of another gate. Once again, he makes a note of the number preceding this one. This is the second number of the combination. Then once again he turns the dial in the reverse direction—the original direction—until the process repeats itself.

After completing this process, and if the intruder has not been tricked by any false gates, he will have the correct numbers of the combination. However, it is by no means certain that he has them in the right order. This is no great problem, though, since he can determine the right sequence simply by varying the sequence of the numbers until he hits the right combination and the lock opens.

The existence of false gates naturally delays this process considerably, but the working principle remains the same. However, lock picking of any kind cannot be learned properly in a short time, and this is even more true for combination lock manipulation. Even experts on combination locks often fail to open a safe in this way (e.g., when the safe's owner has forgotten what combination he chose).

There should always be at least a certain number of divisions on the dial between the different numbers of a genuine combination. This is to avoid any possible malfunctioning within the combination lock mechanism. This means that if the intruder has determined two numbers that are too close to each other or to zero, then he will know that one of them is almost certainly false.

One important factor that assists the manipulator should be mentioned here. It is very common that the safe's owner does not realize that the dial must be revolved completely a number of times (four times for a three-wheel lock and five times for a four-wheel lock) to lock the safe properly. If the dial is only partly rotated, the lock will not be properly locked: only the last wheel has been disengaged from the unlocked position. Of course, the remaining wheels remain in the unlocked position: this naturally helps the intruder who tries to open the lock. You, the owner of the safe, should not give an intruder this benefit.

The intruder doesn't know whether the lock is locked properly or not. He is also unaware of whether the dial (if the lock was improperly locked) was last turned to the left or right. To find out these important details, he will first *very slowly* turn the dial in either direction to feel whether the driving cam is engaging the wheel or not. If it is, he must reverse the direction of the dial at once without going farther in that direction, or he will lock the lock. If he can turn the dial in the opposite direction without the driving cam engaging the wheel, the lock is not locked correctly.

When he has determined that the lock is improperly locked, he will move the dial to the point that he, from previous study, knows is the position in which the gate of the cam is aligned opposite the fence. This point is generally in position 5 or 95, but this is different in various types of safes. By depressing the dial and possibly moving it slightly to the left or right, the burglar now attempts to cause the wheel to align itself to open the lock. If the wheel is only slightly disengaged from the unlocked position and if he is lucky, this simple procedure might open the lock.

It is much more common that an improperly locked combination lock has the last wheel completely disengaged. Then the intruder must concentrate on finding the correct number on the dial that will bring this wheel to the unlocked position. This can be done in the regular way described above. Then only one number needs to be found to unlock the safe.

Yet another point to consider is that somebody who is too careless to lock his expensive safe properly will probably use an easily remembered combination. Certain numbers might have some meaning for him, such as his birthday, or he might have used numbers divisible by five or ten. Combination lock manufacturers advise against such combinations, but careless individuals still use them because they are easy to remember. Do not be among the careless.

It might also be worthwhile to search the room for a written note of the combination. Such notes are frequently found in desk drawers, notebooks, or on a nearby calendar. Now you know where you should not write down your combination.

Some combination locks are operated with timer locks that keep the lock effectively closed until a certain time. Such advanced locks are generally only found in bank vaults, however.

CHAPTER 10

Gun Storage

Most people recommend that you store your guns and ammunition in a safe, and some jurisdictions now demand it. Although storage in a safe is by definition safer than in other places, a safe also brings several disadvantages. First, it draws attention to itself. Any opportunist burglar will mark your house as a place where valuables are stored. He, or his partners, is very likely to return later with better equipment. And by breaching your safe, he conveniently finds all your valuables and weapons in one place. No need for the burglar to waste time searching for them.

Second, a safe is expensive and (to be of any use) very heavy. Usually it must be delivered, which means that the shop personnel and delivery people will know where you live and that you have valuables worth stealing. Do not expect the shop owner or delivery service owner to be too careful with customer records. Anybody with access to these records, from the deliveryman to the owner of the shop, could be in a position to sell your address to the wrong people.

Despite these drawbacks, some jurisdictions demand that you buy a safe. If so, what can you do to enhance the security of it? Some good hiding places for safes were described in the previous chapter. You can always consider having two safes, bought in separate shops. Let one stand in the open, as a tempting target for any would-be intruder. Keep some small things in it, but nothing that cannot be replaced. For instance, a gun owner could keep his ammunition there. The other safe, in which you store your important belongings, should be hidden in the floor or wall.

When discussing the secure storage of guns, we should distinguish between gun dealers, who for obvious reasons must have very secure storage facilities for their firearms and ammunition, and ordinary gun owners, who merely have a few firearms in their home. What may be necessary for the one does not automatically suit the needs of the other.

One further point. At home, you may be tempted to keep a handgun near your bed for home-defense purposes. Unfortunately, every year there are several incidents in which children kill or maim themselves or others by playing with handguns they find in bedroom drawers and similar locations. There can be no greater trauma than to have your child accidentally killed by your own gun merely because you forgot to lock the drawer or your child happened to find the key.

For this reason, I very strongly recommend that you keep all firearms separate from their ammunition and safely locked. Very small children are strong enough to pull a trigger but generally are neither strong nor knowledgeable enough to load a weapon. Unless you happen to be the sole witness in a Mafia trial, you will still have time to load your weapon before an intruder enters your bedroom, at least if you have fortified your home and installed a reliable alarm system of one of the types already discussed. *Do not leave loaded weapons where your children might find them.*

CHAPTER 11

The Small Office or Shop

The information in the chapters on safes and gun storage applies also to many small offices and shops. But shop owners have special problems that the ordinary homeowner need not concern himself with. A very serious issue with shops is theft, both by customers and employees. Offices also have problems with employees stealing goods and perusing confidential files (a concern for many physicians and lawyers). Both problems can be addressed by reliable locks and alarm systems.

Unfortunately, many shops also suffer from repeated burglaries. What can you do to protect your merchandise if the burglars are unimpressed by all the ordinary types of security systems and if security or police personnel consistently arrive too late to catch the burglars? The answer may be the smoke generator.

SHOPLIFTER DETECTION SYSTEMS

Alarm systems designed to detect and scare away shoplifters have been in common use for a long time.

All types of detection systems rely on fastening some kind of indicator or tag to each object to be protected. In an electromagnetic system the tag might be a magnetic tape, whereas in a radio frequency system the tag is a special coil on a circuit card. In either case, the alarm will sound if the tag is passed near a specially designed detector.

The radio frequency system is easiest to use because the electromagnetic detector is very clumsy. However, radio tags are always active, which might cause complications. The radio frequency system also sometimes triggers a false alarm if exposed to portable radios. Furthermore, the tag can be rendered inoperative by hiding it in a metallic cover. This nullifies the signal and prevents the alarm from being sounded. For these reasons, neither of these systems is very well regarded today, even though they are still in common use.

The British company Securitag International is currently the major manufacturer of shoplifter detection systems. Its products are very popular because they rely on a completely different technique.

The Securitag system also attaches tags to the goods to be protected, but tags function differently. The detector posts continuously send out a low-frequency signal, which will trigger any tag brought within range of the signal. The tag is then activated and responds by transmitting another signal—this signal will trigger the alarm.

Whenever a tag is taken through the main door to a store, it automatically transmits a signal that is received by the detector posts positioned either near or on either side of the door (Figure 97). Every detector post has a range of about 90 centimeters. Alternatively, a detector loop can be installed around the door, which negates the need for posts. This design negates the possibility of a false alarm resulting from the presence of portable radios or metal objects.

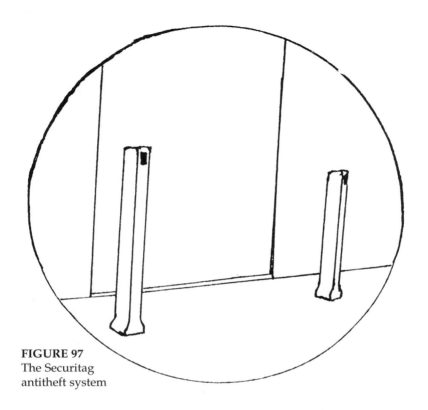

FIGURE 97
The Securitag
antitheft system

The tags are extremely difficult to remove without special equipment from Securitag. However, because this equipment is the same in all Securitag units, an intruder can generally acquire it easily enough should he find it worthwhile to do so. Far more plausible is the threat from ordinary theft.

As a curiosity, it should be mentioned that this idea was originally introduced by the KGB for use in electronic surveillance operations. Similar methods are also known to be used by Western intelligence agencies.

SMOKE GENERATORS

Many shops, especially those selling computer or electronic products, suffer repeated burglaries. Expensive goods are lost, and the insurance costs rise faster than profits can be made.

A few years ago, an English radio shop owner found himself in this situation. Inspired by the artificial smoke used in the nearby theater, he invented a special smoke generator and emitter that he attached to his security system. The effect was that 20 seconds after the alarm had sounded, the entire shop was flooded with dense artificial white smoke. This presented the would-be burglar with a hopeless situation—how could he steal something he could not see?

The smoke emitter became a hit with other shop owners, and a commercial version was soon for sale from Cronwill, England. It or similar systems are now for sale in several countries. The smoke, which actually is a kind of mist, is produced from a mixture of 70 percent glycol and 30 percent distilled water, heated until it turns into dense smoke. The smoke is harmless and does not damage electronics or furniture.

As soon as the alarm is sounded, the smoke is emitted and pumped into the building at high speed. A special sensor, working in much the same way as a smoke detector, controls the amount of smoke so that it becomes as dense as possible. An area 3 x 3 x 20 meters will be flooded in about 20 seconds. The smoke emitter will continue to produce smoke for several hours or until the system is turned off. In fact, the smoke is so dense that in a major warehouse, the burglar may not even be able to find his way out before the police arrive. After the burglary attempt, the smoke disappears naturally as soon as the smoke emitter is turned off, with no need to clean up the building or merchandise (it also dissolves in case of fire, so as not to hamper the work of the firemen). Furthermore, modifications of the basic system are under way in Britain to use the smoke to mark the burglar with ultraviolet paint for quick identification and to provide evidence that can be used to prosecute him.

The smoke emitter should be aimed low because the smoke will rise toward the ceiling. It is best to mount the emitter above the ceiling but to aim it downward (e.g., through a ventilation shaft). To keep the smoke generator hidden and safe from sabotage, you can install a number of dummy shafts.

Smoke emitters of this kind are also available for residential homes. Although not cheap, the device is a reliable means to prevent burglars from stealing your belongings. Do invest in such a system if you can afford it, but don't forget that the smoke system is only a means to discourage the burglar and prevent his work. You must still have a reliable system of alarm sensors to first detect the burglary.

THE SMALL OFFICE OR SHOP

CHAPTER 12

Vehicle Security

In recent years auto theft has grown into a highly sophisticated, organized criminal activity. Not only are expensive luxury cars stolen to be resold (often in another country), popular medium-priced cars are also a favorite target because their parts are easily resold on the black market. In Great Britain, police statistics indicate that one car is broken into every 20 seconds. Theft of and from cars is one and a half times more common than house burglary. In the United States, too, vehicles are frequent targets for criminals. Like house burglaries, roughly 80 percent of all thefts involving cars are committed by opportunist thieves. Too many people forget to lock a door or close a window completely, especially if they plan to be away for only a minute or two.

Even if you remember to lock all doors to your vehicle and roll up the windows, cars are very easy to break into. Thieves look for briefcases or bags left in the car, and if these are in full view, as is often the case, the vehicle can be entered in a few seconds. Usually the intruder has more time at his disposal, and he knows it: the thief simply checks the pay-as-you-leave parking slip you left in the car or the amount of time left on the parking meter.

If you park in a multistory parking garage, keep the car park ticket with you. Thieves have been known to break into cars and then use the owner's ticket to drive out. Additionally, try to park in the busiest part of the garage and avoid the floors farthest from the attendant's office. That way, you can also avoid all those dark stairways and unpleasant elevators. In an outdoor parking lot, park in the middle rather than round the edges where trees or bushes may provide hiding places for thieves to lurk.

Never leave valuables in your car. In fact, it is best not to leave any object—valuable or not—to invite a thief. Favorite hunting grounds for car thieves include shopping mall parking lots and service stations located near exit ramps, where many people leave luggage in their cars while they have a snack or meal. Some thieves even siphon gas out of tanks that don't lock, and car wheels often get stolen too.

Although empty parking lots may be unpleasant, you will be most at risk the moment that you actually lock or unlock your car. If you see suspicious people loitering nearby, do not get into your car; retrace your steps to a well-lit area where there is an attendant or someone else you can alert.

Your home security system should also include measures to protect your vehicle. Do not spend too much effort on this, however, because it is almost impossible to fully secure the average car. Do use basic precautions, as detailed below, but consider any car theft an insurance matter.

VEHICLE LOCKS

Vehicle locks can be of almost any type, including the previously described disk tumbler and pin tumbler locks. But vehicle locks often rely on the sidebar principle. The sidebar lock— in fact, a variety of the disk tumbler lock—is a more specialized vehicle lock. This lock is commonly used for ignition, door, and trunk locks.

The sidebar lock is fairly simple in construction (Figure 98). There are disk tumblers inside the lock, with V-shaped notches in their sides. When the proper key is inserted and engages the tumblers, the key aligns them so that the spring-loaded sidebar moves out of the cylinder and into the plug. The plug is free to rotate when the sidebar has passed the shear line. This will unlock the lock.

Although the sidebar lock is simple in construction, the sidebar makes the lock difficult and time consuming to pick. There is no way to determine when the sidebar will fall in place, because it is impossible to hear or feel the tumblers align with the shear line. Here experience is necessary, unless the thief opts to drill the lock open. Then an L-

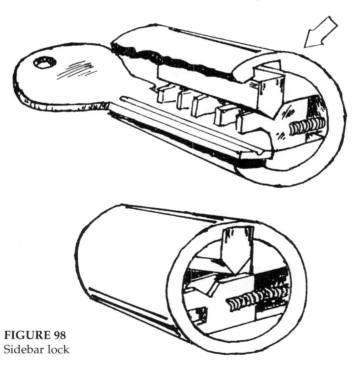

FIGURE 98
Sidebar lock

shaped wire can be used to put pressure on the sidebar, while the thief rakes the disk tumblers into place. Such a method leaves very clear marks indicating that the lock has been tampered with.

The sidebar principle is also used on certain pin tumbler cylinders. Sidebars are found in high-security locks only.

VEHICLE DOORS, WINDOWS, AND TRUNKS

There are numerous ways of entering the average automobile. The door windows, front and rear ventilation windows, doors, and trunk are all possible entrances. Sometimes, the locks are easy to pick; at other times it may be easier to employ some other method of entry. In this section we will look at some of these other methods used by car thieves. (The methods burglars use to pick the locks were described in an earlier chapter.)

The most vulnerable parts of an automobile are usually the windows. One reason for this is that most windows allow easy reach to the door-release push-button lever. This is especially true for the contemporary models that have a single pane of glass. A thief can often easily reach and lift the push button with the help of a bent coat hanger. The coat hanger is first straightened out and then the end is bent into a loop or triangle (Figure 99).

First, because the window is usually rolled up tight, the thief will force a paint scraper or a similar object between the

FIGURE 99
The coat hanger bent in a loop or triangle

edge of the window and the weather-stripping (Figure 100). This creates an opening large enough to allow the coat hanger to be inserted. The push-button lever can then be caught by the loop or triangle and pulled up to open the door.

Another possibility is to use a gun-cleaning rod instead of a coat hanger. The gun-cleaning rod comes in sections, so it is easy to make the rod long enough to reach across the inside of the car. In this case, the thief always works on the window or door opposite the side where he inserts the rod.

First, the thief makes a loop with some nylon line and fixes it to the end of the rod on the slotted cleaning attachment on the end of the rod (Figure 101).

FIGURE 100
The paint scraper forced between the edge of the window and weather-stripping

Then he inserts the rod and catches the lever or the door handle inside the car with the loop. He pulls on the line and lifts the rod to raise the lever or the handle caught in the loop to the open position.

If the car has a rear ventilation (or wing) window, it can also be used to gain entrance to the car. These windows come in two types, those with and those without a locking button on the swivel. If there is no locking button, it is easy to force the swivel lock up and into the unlocked position. A thief will simply insert a paint scraper or a similar object between the window and the frame. Then he will bend the tool slightly to make an opening wide enough to allow a thin piece of wire to enter. He will take care to loop this wire around the swivel level and then pull it upward. This is most easily accomplished if the wire is slightly bent on the end, so that it does not slip off the lever.

Most modern cars have locking push buttons in addition to the swivel levers, but there are special tools available to take care of them as well. Two tools are required, and they are

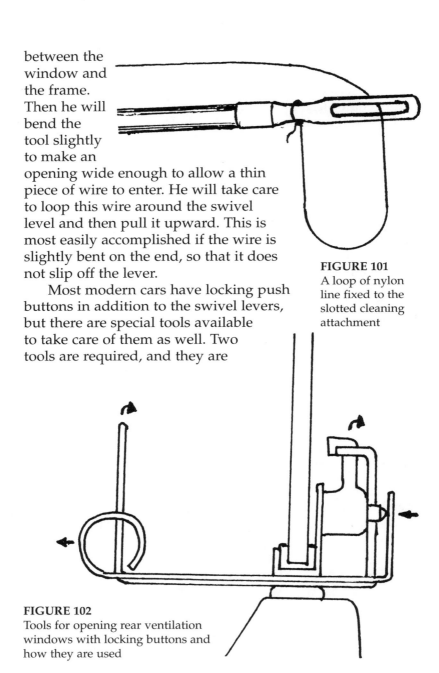

FIGURE 101
A loop of nylon line fixed to the slotted cleaning attachment

FIGURE 102
Tools for opening rear ventilation windows with locking buttons and how they are used

inserted on different sides of the lock. The first tool is used to depress the button by pulling the tool toward the thief, while the second one is twisted slightly to push the lever into the unlocked position (Figure 102).

The front ventilation window is another popular way of gaining entry to the vehicle. This allows access to both the door handle and the window roller handle.

Here, too, special tools are available and very simple to use. A thief pries the window open slightly, inserts the tool, and turns the handle (Figure 103). The tool chosen generally depends on the amount of working space available since the function is the same.

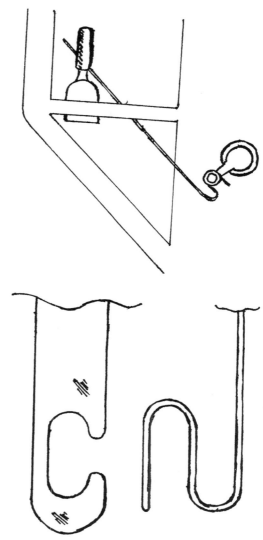

FIGURE 103
Opening the front ventilation window and tools for this purpose

VEHICLE SECURITY

The trunk of the car is a completely different matter. Since the same key is often used both for the door locks, the ignition, the trunk, and even the glove compartment if it is locked, all these locks can be dealt with at the same time. Picking is the most common method to open these locks, but it is too time consuming for the average thief. Some choose to drill the lock open, but many manufacturers have protected their locks by installing various steel plates or pins in front of the catch to make drilling more difficult. Besides, if drilling is required, it is often easier to break the trunk open or to drill beside the catch lock and then to manipulate the catch with a bent piece of wire. In either case, the intrusion will naturally leave very clear marks on the car.

A few trunk or tailgate cylinder locks, as well as many glove compartment locks, can be opened in a much easier way. If you are concerned about car theft, examine the lock to see if this method is possible in your car.

A simple lock, possible to open in this way, is designed to be secured by a retainer accessible from the front of the lock. This lock can be opened by inserting a special L-shaped tool, in effect a 5- or 6-millimeter hook in the end of a piece of stiff wire, through the keyway (Figure 104). The retainer can then be pulled down and worked free. Next, the entire plug can be pulled out of the lock and removed. To do this, the thief forces the retainer (usually installed with its open ends toward the passenger door) toward the center of the car to disengage it. When the plug has been extracted, the catch mechanism can be pushed back and released with the help of any pointed tool.

There are in fact two different locks that can be opened with such an L-shaped wire. The lock in which the plug can be removed has been described already. The other lock is of an even simpler construction. The wire can, when inserted into the keyway, engage the catch mechanism itself and open it by merely pushing it downward.

Today's vehicles are protected in many different ways. In some cars, for instance, the standard locks have been replaced

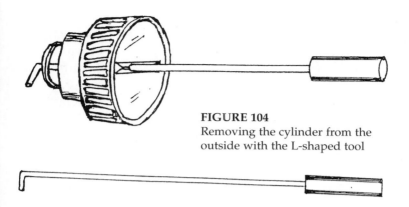

FIGURE 104
Removing the cylinder from the outside with the L-shaped tool

by new ones. Still, because too many people frequently drop their car keys, some cars have a magnetic box containing a spare key that fits under the car. If the burglar finds such a box, he just opens the box and retrieves the key.

CAR ALARM SYSTEMS

There are a lot of different car alarm systems available on the market today. Some of them sound the alarm if somebody tries to jimmy a door or the trunk of the vehicle. Other alarm systems trigger the alarm if someone attempts to move the car (e.g., by towing it away). Certain alarm systems even include a remote pager to signal the owner if his car is being tampered with.

As in ordinary alarm systems, the car alarm includes a control unit, one or more sensors, and a warning device. In a car alarm, however, some of these devices—notably the control unit—are simpler in design than the units available for protection of homes and office buildings.

In many cases, the car alarm is at least partly self-contained. No self-contained alarm system is able to protect the entire car, including its trunk and hood.

The control unit is often mounted near or on the dashboard, either by using a mounting bracket similar to those used for radios or stereo systems or by mounting it directly to the surface. Because an owner generally does not want to give a potential intruder the opportunity to see the control unit, it is often hidden under the dashboard, under a seat, along the fire wall, or in the glove compartment. A backup battery is often provided, too, to increase the reliability of the system.

Most types of car alarms are operated by either a concealed switch or a key. The sensors used in car alarm systems can be divided roughly into two types: pin switches, guarding the entry points in the vehicle, and different types of motion detectors designed to detect movement of the vehicle.

The most common type of car alarm sensor is the pin switch (Figure 105), also known as the earth-seeking sensor. This is a door-contact switch that is activated when the electric current is broken. A pin switch is wired directly to the door, installed in the framing around the doors and/or the trunk and the engine compartment. Pin switches can also be used to guard the trunk, hood, sunroof, tailgate, or other points of entry.

Pin switches are spring-loaded, momentary-contact switches similar to the plunger switches described in a previous chapter. The pin is released from the switch when the door is opened, which electrically grounds the system, thus triggering the alarm. This device is either linked to the car horn or to an independent siren. An independent siren is a better option because it is more difficult to find and disconnect.

Since pin switches have only a single wire connected to them—the car's metal chassis being used as the ground portion of the circuit—they are always installed on a metal surface. When used to protect a door, the pin switch is installed near the switch for the interior light on the lower part of the door post.

Some car alarm systems use the switches already installed in the door frames of the car instead of pin switches. These already installed switches are used to turn on the

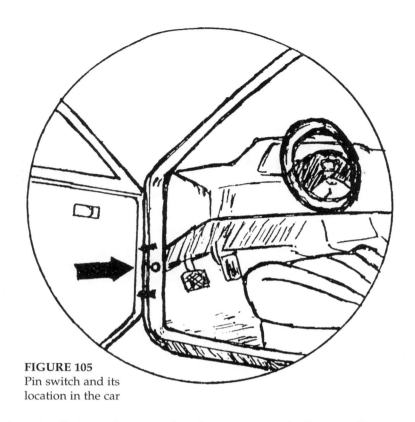

FIGURE 105
Pin switch and its location in the car

interior lights whenever the door is opened. Electrically, these switches are identical to pin switches. In this case, the door frame switches are added to the alarm system by attaching wires to them from the control unit.

The other type of car alarm, in very frequent use but not really very reliable, is the motion detector. This is a special vibration switch designed to sound the alarm when the car is shaken or moved. This sensor, although popular, has a tendency to give false alarms but is very widely encountered. Some control units incorporate such a sensor in their housing, to protect the control unit itself as well as the car. Such a self-contained car alarm is difficult to remove without triggering the warning device.

If a motion detector is integrated into the control unit, the unit must always be as level as possible, preferably near the center of the car. If positioned in some other location, it will not be as sensitive to motion at the front or back of the vehicle. The most common location is under the dashboard.

Among the many types of motion detectors is the pendulum alarm. This alarm has a pendulum switch that is used to sense vibration or motion. The switch is designed with a set of contacts that come into contact with each other when the switch is moved or shaken. It is set off when the car is rocked, jolted, or otherwise moved (e.g., when being towed away).

The pendulum itself is a small weight on the end of a light spring with a contact underneath the weight. Whenever something causes the spring to vibrate, the weight touches the contact and the alarm is sounded. The sensitivity of the spring can be adjusted, but most car owners do not do this. The result is that the pendulum alarm can be triggered by almost anything, such as the vibration of a passing truck or somebody who accidentally bumps into the car. Even a very light jolt is enough to trigger this alarm. This propensity for false alarms has made the pendulum alarm almost useless. In neighborhoods where this type of alarm is in widespread use, no one thinks twice when hearing it, and there is usually no response to it.

A slightly more advanced version of the pendulum alarm is the trembler switch alarm. This device is similar to the pendulum alarm but, instead, has a ball bearing sitting between two contacts. Any definite movement of the car, such as somebody's trying to open a door, will cause the ball-bearing to touch the contacts and trigger the alarm.

Both the pendulum and the trembler switch alarms are generally mounted under the hood, on the fire wall in the engine compartment, and can be adjusted for sensitivity. Another common place for these motion detectors is under the dashboard.

The vibrator contact circuit, or adjustable impact device, is another motion detector. It has a relatively light weight on the end of a piece of spring steel, with a contact under the weight. When the spring steel vibrates, the weight touches the contact, thus activating the alarm. In this device, sensitivity can be controlled by adjusting the distance between the weight and the contact.

Yet another type of motion detector is the microtransducer (Figure 106). A transducer is a device that produces an electric current in response to vibration, shock, or motion. This is a very small sensor, generally round, flat, and shaped like a small coin. It has a small interior piezoelectric crystal, which is affected by vibration, very loud noises, or the car's being hit by something. A piezoelectric crystal is a crystalline material that develops a voltage when subjected to such mechanical stress or severe vibration. When vibrating, this sensor produces an electric current, which in turn triggers the alarm.

Microtransducers are often located inside the door frame, near the switch for the interior light, in the same position you would expect to find a pin switch. Microtransducers, however, can also be found mounted on the center door posts of four-door vehicles. In either case, they are glued into place with epoxy.

Self-contained vibration sensors are also sometimes used. Such a sensor consists simply of a keypad fitted to the dashboard. The alarm is activated by entering a personal code and works just like any other motion detector.

Another car alarm sensor is the voltage drop sensor, also known as the current-drain sensor. In this device, a sensor is wired into the electrical circuit of the car. Whenever the sensor detects a drop in the voltage (e.g., when the courtesy light comes on as the door is opened or the key is put in the ignition and the dashboard lights up), the device will trigger the alarm. This

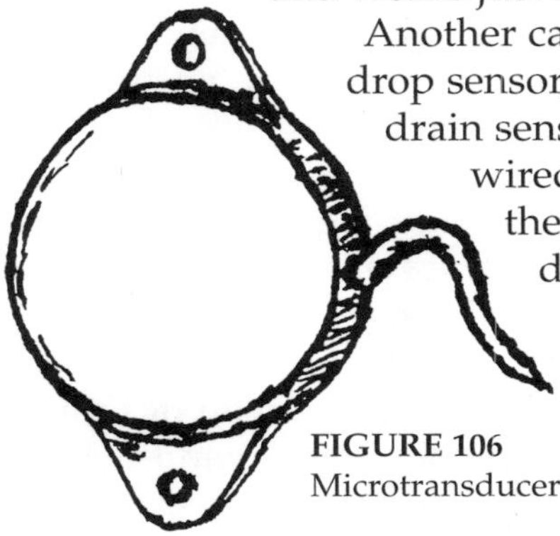

FIGURE 106
Microtransducer

type of alarm can easily be disconnected, accidentally or not (e.g., by the courtesy light's failing for some reason) and is therefore not very reliable.

More advanced types of alarms are also used as car alarms. Among these is the ultrasonic detector (described in a previous chapter), the most common location for which is on the shelf behind the rear seats. Do not install this detector in the dashboard, because it can be blinded by a thief's opening a side window and slowly moving a sheet of paper in front of the detector, as described earlier.

A siren is the best choice for a warning device, and various kinds are commonly used today. They are often mounted under the hood, although well away from sources of extreme heat. The siren is generally mounted slightly downward to prevent excessive accumulation of dirt and moisture, and it is almost always connected only to the control unit. As always, this connection is the vulnerable link in the alarm system.

Often the siren will be supplemented by other features, such as a light-flashing facility. This device can easily be wired into any of the many types of alarm systems. Finally, remote sensor alarms can be fitted to other areas, such as roof racks and trailers. The really security-conscious can mix all these options in the same alarm system.

Other specialized car alarm features include the "passive" alarm system, which automatically arms itself once the owner has locked the car. This alarm is often armed and disarmed by means of a concealed switch inside the car. Therefore, a certain delay is imposed before the alarm is sounded, so that the owner will have enough time to open the car door and disconnect the alarm system. Other alarm systems are manually armed and disarmed through the use of an external security key switch, usually located at the rear of the vehicle. A variant of this system relies on an IR transmitter, kept on the owner's key ring or another hand-held device. After stepping out from the car, the owner simply aims the device toward the receiver, mounted inside the car, and presses a button. The IR beam activates the system, locking all doors and setting the alarm. In some versions of this alarm system, the car flashes its headlights to indicate that the message was received and understood. The alarm can be switched off and the doors unlocked in the same way. Another variant of the same idea is the key ring containing a small radio transmitter that emits a radio signal. Locks of these types are generally called remote-control locks.

An interesting option on many control units is a built-in radio transceiver (a receiver with transmitter). This device allows the alarm system to be armed and disarmed, or even tripped, by a miniature radio transmitter built into a small pager unit. This circuit alerts the owner to the fact that his car is being tampered with.

Remote paging car alarm systems of this type are especially popular in areas where there is little likelihood of somebody else noticing the alarm if it is sounded. The owner carries a remote paging receiver with him, so he can always be alerted when the alarm is sounded. He can then investigate by himself, call the police, or do both.

These systems are sometimes employed, without the use of an ordinary siren, as a silent alarm. This is especially likely if the owner hopes to catch the intruder in the act. Some alarms of this type allow the user to choose between siren and silent alarm operation.

Those remote paging systems that are able to arm or disarm the alarm system from a distance almost invariably include a panic button, with which the owner can sound the car's siren if he is threatened or surprised by intruders. This option is popular both among people afraid to walk alone through empty parking lots and professional truck drivers, who might have to stay for long periods around their vehicles and also frequently sleep in them.

The remote paging system consists of two parts. The combined transmitter and control unit is mounted in the car and powered by the vehicle's electrical system, whereas the remote paging receiver is small enough to keep in one's pocket and is battery powered. Although these two devices work on a radio channel on a frequency away from most other radios, they are also safeguarded by a security code sent when the transmitter is triggered. This prevents accidental triggering and is also supposed to preclude an intruder from using his own transmitter to disarm the alarm system. Of course, the latter is not true, since an intruder can easily determine, and imitate, this code if he has access to specialized equipment. Such equipment is costly, however, and not in common use among car thieves.

When the transmitter is triggered, it sends out a signal containing the security code. The paging unit, when detecting a signal on its preset frequency with the proper code, starts beeping. It generally also flashes an LED. The security code is selected in advance by setting a group of internal dip switches, which can be found in both devices.

The transmitter uses the vehicle's standard antenna, including the type that automatically rises when the radio is switched on. Therefore, the intruder can keep the antenna from rising, thus severely decreasing the range of the transmission. The vehicle might also have a wire-type antenna built into the windshield, but the effective range of the transmitter will be reduced significantly in any case. Some vehicle owners attempt to extend the range of the transmitter by installing a separate antenna. Remember, though, that the transmission in this alarm system functions as the wiring in ordinary alarm systems and is therefore a prime target for the intruder, who will try to disable it. If the alarm-call transmission can be prevented, the owner will not be alerted.

Generally, though, the transmission will be sent. The range is then completely dependent on the terrain and the characteristics of the surrounding area. The range might be several kilometers in open country but reduced to a few hundred meters, or less, in a city or in an underground parking garage. Radio transmissions are also often reduced considerably in strength by the metal used in most high-rise buildings. Most alarm systems manufacturers claim an "average" range of 3 kilometers, but this is quite an exaggeration. Between 200 and 500 meters is more typical in a city.

The transmission can be completely eliminated by enclosing the transmitter in a metal box. The car, in effect, would be such a box, if it were not for the antenna and, to a lesser degree, the windows.

Many car alarms are modified for use in other vehicles and locations, such as boats, trailers, or campers. Remote paging systems are especially popular, because they have a great range (moreso if connected to a citizen's band base station antenna). They are sometimes used to guard scattered buildings or stores on farms and construction sites.

Arming and disarming the car alarm is otherwise performed by several different methods, also depending on whether a delay is built into the system or not. If so, the delay might be as short as 12 seconds or as long as 40 seconds or more. Arming can, for instance, be done by turning the ignition key switch to the ON or ACC (accessory) position for a few seconds and then switching it to OFF. This arms the alarm system after an exit delay period. Disarming is then done by simply entering the car and switching the ignition key switch to ON before the entry delay period is up.

Other arming and disarming methods include switches on the control unit. The easiest method is when the process is performed automatically whenever the ignition is switched on or off. In many cases, the control unit informs the user of the fact that it is armed or disarmed by producing a beep, emitting an LED, or sometimes both.

A valet switch might also be present. This switch allows the owner to bypass the alarm system in situations where he expects to be away from his car, but the car will be attended or guarded (e.g., by valet parking or servicing). The valet switch can only be activated while the engine is running and is of little use to an intruder.

Many car alarms are powered by the car's own battery, but it is also quite possible to fit a second power unit as a backup. The latter is done mainly in commercial vehicles because in many of these vehicles it is easy to gain access to the battery terminals and disconnect the ordinary alarm. Remember that it might be possible to disconnect the wiring from underneath even if the battery is locked under the hood. A backup power facility is always a prudent acquisition.

Almost all car alarm systems are designed to be powered by 12-volt DC current. Furthermore, the wiring is often easy to identify, since a color-coding scheme is in common use in many countries. For instance, red wiring might be used to connect the system to the power source, the car battery, and other colors might identify the components of other particular systems. Remember, though, that this often varies from country to country and sometimes among manufacturers in the same country.

A professional car alarm system also has closed-circuit wiring. This means that the alarm will sound even if the wires to the sensors are cut. Despite this, however, the system will not function if the wire to the warning device is cut instead. So, again, we see an expensive system recommended by the professionals that can be disabled very easily.

Sometimes the purpose of the break-in is not to steal items from the vehicle, but the vehicle itself. This leads to several other considerations, apart from the lock and alarm system. Although you generally cannot protect yourself from somebody really intent on breaking into your vehicle, there are efficient ways, both electrical and mechanical, to immobilize the car so that it cannot be stolen.

The electrical means of immobilization include fitting one or more devices to the car, such as an ignition cut-out device. This can be a part of the vehicular alarm system and is linked either to the car's horn or any other siren being used. The device sounds a warning at the same time that it blocks the ignition circuit, automatically immobilizing the car.

A manual ignition cut-out device is also useful. An ignition disabler switch that can be hidden under the dashboard will interrupt the ignition feed wire. Note that the car can still easily be hot-wired and started if the interruption is made between the battery and the coil. If the interruption is made between the coil and the distributor, however, hot-wiring is not usually possible. The switch might be locked and require the use of a key to open.

A passively armed cut-out device renders the ignition dead as soon as it is switched off. When the driver wants to start the car again, he must deactivate the cut-out by depressing a button (or, alternatively, switch) while he starts the engine.

Yet another method is to use a removable circuit card. Usually this card is put into a socket mounted on the dashboard. When the card is removed, vital electrical circuits are broken and the car is prevented from starting.

One method of electrical immobilization is to fit a switch that interrupts the feed wire to the electric fuel pump. Or a multiple cut-out device can also be incorporated into the central control unit, which is bolted to the bulkhead under the hood that disrupts several electrical circuits at the same time. Finally, another, far simpler electrical method of immobilizing the vehicle is simply to swap or remove a couple of the spark plug leads.

Mechanical immobilization methods include fitting an engine immobilizer switch, the previously mentioned hidden switch in the ignition circuit, or an internal locking device (e.g., fitting over the hand brake and locking around the gear lever). Simpler locks, but also fairly reliable, include combination locks attached to the hand brake, engaged or disengaged by means of a three-digit combination. Such locks slide over the top of the hand-brake lever, locking it in the "on" position. Electronic locks that prevent the engine from starting until the driver has entered the correct code on a keypad fixed to the dashboard can also be found.

Additional steering wheel locks are also commercially available, although they often are of the hook type, hooking on or over the steering wheel and brake, accelerator, clutch pedal (Figure 107), or floor-mounted gear stick. An intruder can get rid of this device if it is hooked over the clutch pedal simply by stamping down hard on the clutch. Since the steering wheel bends quite easily, the lock will come off. If the brake pedal is used instead of the clutch pedal, however, this procedure is sometimes more difficult to perform. Of course, the quality of the locking device will also affect the outcome.

Additional locks can be fitted around the steering column like an armored collar. The lock key then replaces the vehicle ignition key and controls the electrical operations. Although in principle this is a good lock, a thief can demolish it with heavy-duty tools.

Some wary individuals remove the rotor from the distributor to protect their cars from theft. The distributor cap is easy and quick to snap off to remove the T-shaped rotor sitting in the middle of the distributor. It is small enough to put in a handbag or pocket. Others remove it and then lock it in the trunk. A final, and very definite, method of mechanical immobilization is to use a wheel clamp, which effectively prevents the car from being moved. Most wheel clamps also prevent the tire and wheel

FIGURE 107
Steering wheel hook lock

HOME SECURITY

from being removed. Such clamps can be found on cars that are left unattended for considerable periods of time or that have been secured by the police.

Many expensive cars today have additional electronic security devices. BMW uses a mechanical key with a built-in chip carrying electronic coding and a code that changes each time the driver starts the engine. This renders the engine impossible to start by merely connecting the ignition wires manually.

Campers, trailers, and motorcycles present other problems. Campers and trailers are sometimes safeguarded from being towed away by locking a hitch lock, or towball, into the ball socket of the vehicle-towing hitch. Such a device can only be removed by using the correct key bar. A thief must pick the lock or break it.

Motorcycles and bicycles are generally secured with chain locks. The chains are generally easy to cut through with bolt cutters, especially if they are not of hardened steel. For this reason, you should use a specially designed padlock with hardened, elongated shackles. The steering lock of a motorcycle is usually of simple construction and can often be broken by a fierce wrenching of the handlebars. Although motorcycles can be fitted with electrical immobilization devices and alarm systems, this is very uncommon. It is more usual is to immobilize the motorcycle by removing the battery ground strap or the line fuse in the main lead near the battery terminals. A concealed cut-out switch that breaks any of the low-tension wires to the coil can also be fitted. As long as the correct equipment is available, however, neither method presents any problems to the thief.

Other measures to protect a motorcycle for long periods include such devious alterations as fitting unserviceable but visibly complete spark plugs, draining the float chambers and removing or blocking the fuel supply line, or selecting first gear and then removing the gear and clutch levers. Once again, this is only a matter of having the correct spares available should a thief really need to move that motorcycle. As with other vehicles, there are no completely reliable ways to safeguard a motorcycle from theft.

Never leave your crash helmet with the motorcycle. Where it is illegal to ride without helmet, a thief does not want to attract police attention by doing so.

CHAPTER 13
Personal Safety at Home

Although taking adequate mechanical and electronic security measures to protect your home and vehicles is important, you and your family must also make sure that you do not unnecessarily expose yourselves to personal danger. This is obviously a matter of personal choice, but in this chapter a few guidelines will be offered, most being concerned with the effects of burglary.

When you arrive at home, always have your house keys ready to let yourself in without delay or fumbling. This reduces the chance of attack by opportunist muggers.

If your arrive home to find that a burglary has taken place, do not go in if you think the intruder might still be there. An open front door is a sign that the burglar might still be in your house. Leave quickly and quietly and contact the police—unless you are utterly confident that you can handle one or several armed, drug-crazed maniacs. If you see the burglar, try to get a good look at him and his vehicle, if any. Note the registration number, color, and make of his car, as well as any distinctive marks. Do not let him notice you if you can avoid it.

If your are sure that the burglar has left, do not touch or try to put things back in place until you have been assured by the police that the evidence collection is finished. If you are alone, ask a friend or neighbor to stay with you.

When you have finished with the police, arrange for immediate repairs to broken windows, locks, or doors. If doors or windows need to be replaced or repaired (and here you don't have to wait to be burglarized), use the opportunity to improve your security. Some burglars do come back, especially if they found a safe in your home that they didn't have the time or tools on hand to crack.

Immediately report the loss of any stolen savings passbooks, checkbooks, credit cards, vehicle registrations, and other important documents to the relevant authorities. Also notify your insurance company that you have been burglarized.

Do not underestimate the psychological reaction of being the victim of a burglary. Many victims suffer from shivering and shaking, insomnia, and a detached or dazed feeling that can last for weeks. A few, many of them women, suffer acute reactions, including vomiting or hysteria.

You should also take care if you suddenly, perhaps at night, hear an intruder while you are in your home. It is safest is to move about and make a lot of noise. Switch lights on and off and call out to an (imaginary, if necessary) male companion ("Peter! What was that noise?"). Call the police from your bedroom. Most burglars will flee empty-handed rather than confront you.

We have already discussed the weaknesses of certain door chains. A strong door chain or limiter is very useful, but you should not leave it on at all times—if you suffer an accident at home somebody will need to get in to help you. You should also consider installing a door viewer with a fish-eye lens that gives you a 180-degree field of vision. For the viewer to be useful at night, you need a porch light. If you live in a block of flats where there is no light in the passageways except those that come with a push-button timer switch, always insist that a caller press the timer switch so that you can get a good look at him through your door viewer before you open the door.

If your children need to be left at home alone, warn them not to answer the door under any circumstances or let any stranger in. They must also be warned not to tell any stranger on the telephone that they are at home alone. Make sure they know how to dial you, a neighbor, or the police in an emergency. If you employ a baby-sitter, make sure that you know him or her, including personal/professional background and where he or she lives. Make sure that the sitter does not ask somebody you don't know to take his or her place at the last moment. Also telephone home during the evening to make sure the sitter is actually there. Ask your children, in a casual way, what they think of a particular sitter before you use him or her again.

CHAPTER 14

Security Away from Home

Travel has always been a source of additional danger—and for burglary too. This remains true even today. This book cannot attempt to detail every danger that you may be exposed to on a journey; nonetheless, a few words on common dangers and suitable precautions are offered.

First, don't forget that your home is at even greater risk when you are not there to look after it. If possible, employ a house-sitter to live in your house in your absence. If this isn't feasible, do everything possible to disguise your absence and make your home secure. Cancel newspaper deliveries and ask a neighbor or friend to collect mail, mow the grass, park in your driveway, or move your car from time to time. During the winter, you will have to ask somebody to clear away snow and make wheel tracks in the snow next to your garage. To any opportunist burglar, the house should present a lived-in look.

Professional burglars often frequent airports and similar points of departure looking at labels for the home addresses of people going away. If your bag displays your home address, and if the bag or your appearance gives an impression of sufficient wealth, the burglar may decide to pay you a visit while you are away.

Airports, train stations, bus terminals, and hotel lobbies are favorite hunting grounds for thieves and pickpockets. Do not let your luggage out of your sight, not even for the moment it takes to make a telephone call, use a rest room, or make a purchase. Professional thieves work in teams. While one distracts you; another steals your purse, briefcase, or suitcase; and a third will whisk away the stolen goods. In some countries, many pickpockets are children who swarm around, talking all at once and pulling on your clothing to distract you so they can get your valuables.

If you fear that you will fall asleep while in an airport waiting area or lounge, use your handbag or briefcase as pillow or make certain that no thief can take it without awakening you.

When you have checked into any hotel or boarded any airplane, immediately familiarize yourself with the fire exits nearest your room or seat. Remember that fire is often accompanied by power failures and dense smoke, and you are in unfamiliar territory. Never use an elevator.

In your hotel, you should also check that windows and balcony doors are locked so that nobody can gain entrance to your room at night when you are asleep or while you are out. Keep curtains drawn after dark and when you are out.

When a burglar enters a hotel room, he will search below the wall-to-wall carpet, in the hotel room Bible, among your dirty laundry and luggage, in the pillow, in the curtain hem, in the minibar, behind the sink, behind the bath tub, and below the bathroom carpet. Do not hide your cash in any of these locations.

Many hotels have installed safety deposit boxes in the wardrobes of each room. These are locked with a special key or an individual code. As long as the safety deposit box is correctly secured and cannot be removed, even with force, these boxes are reasonably safe places to store cash, passports, traveler's checks, tickets, jewelry, and other valuables. Most hotels also have safety deposit boxes near the front desk. However, burglars know about these and raid them often. The hotel generally refuses to assume responsibility, so one often wonders how involved the hotel staff is in the crime. Besides, the central deposit boxes are also inconvenient and often not accessible 24 hours a day. Use them only if you have no other choice.

Before you leave home, take a few minutes to consider what documents you will need for the trip. Never bring documents unrelated to your journey, such as local charge cards and library cards. Record in a notebook or on a piece of paper (that you will carry in a separate location) the numbers of all credit cards and documents (e.g., passport, driver's license) you will be taking with you. Do the same with the serial numbers of any traveler's checks. For international travel, it is strongly recommended that you also carry a photocopy of the pages of your passport showing your photograph, name, signature, and passport number. If your passport is lost or stolen, you can use this copy to prove your credentials as well as to get an emergency passport from your embassy without the normal hassle and delays associated with such a request.

Preferably, you should carry the notes and photocopies with your spare currency in a money belt worn under your clothes. Carry only as much cash in your wallet as you expect to use in one day. Even better is to keep some loose cash in your shirt pocket, so that you do not need to show your wallet for very small purchases. Expect pickpockets to empty your pockets or handbag at least once in your lifetime—and at the time when you least can afford it. Another hazard is motor scooter thieves. They work in pairs: one drives and the other grabs your bag as they speed past.

Foreign locations are generally no more dangerous than at home. Take the same precautions in foreign countries that you would take at home. Do not be afraid, but be aware. And enjoy your trip.